ONE PICTURE TO
UNDERSTAND PATENT
EXAMINING PROCEDURE

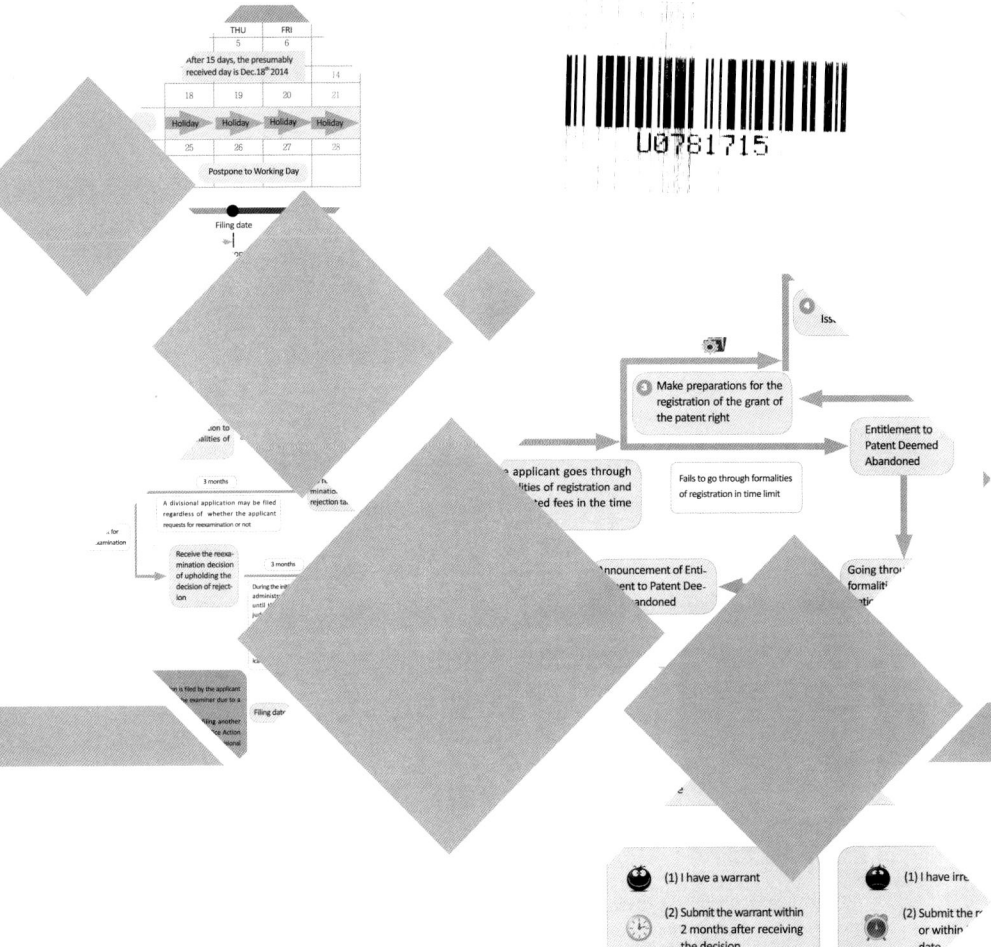

一张图看懂
专利审批流程
（汉英对照）

国家知识产权局专利局初审及流程管理部 ◎ 编

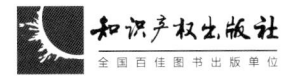

知识产权出版社

全国百佳图书出版单位

图书在版编目（CIP）数据

一张图看懂专利审批流程：汉英对照 / 国家知识产权局专利局初审及流程管理部编 .—北京：知识产权出版社，2017.9

ISBN 978-7-5130-5093-7

Ⅰ. ①一… Ⅱ. ①国… Ⅲ. ①专利制度—审批制度—流程—图解 Ⅳ. ① G306.3-64

中国版本图书馆 CIP 数据核字（2017）第 211501 号

内容提要

　　本书以中英文两个版本共76张图片，简洁、直观地解析专利审查流程中的业务。本书首先向读者概要展示专利申请审查流程，再按照流程顺序，依次展示受理阶段、发明专利初步审查阶段、PCT 申请国际及国家阶段、办理登记手续和专利权维持阶段的主要手续，以及相关的流程服务和行政复议程序，内容基本实现专利审查流程的全覆盖。本书是编者为广大创新主体提供便捷、有效的专利审查流程服务，为社会各界人士快速了解相关专利审查流程的呕心力作。

责任编辑：卢海鹰　胡文彬　　　　　　　　**责任校对**：王　岩

封面设计：张　冀　　　　　　　　　　　　**责任出版**：刘译文

一张图看懂专利审批流程（汉英对照）

国家知识产权局专利局初审及流程管理部　编

出版发行：知识产权出版社有限责任公司	**网　　址**：http://www.ipph.cn
社　　址：北京市海淀区气象路 50 号院	**邮　　编**：100081
责编电话：010-82000860 转 8031	**责编邮箱**：huwenbin@cnipr.com
发行电话：010-82000860 转 8101/8102	**发行传真**：010-82000893/82005070/82000270
印　　刷：北京嘉恒彩色印刷有限责任公司	**经　　销**：各大网上书店、新华书店及相关专业书店
开　　本：889mm×1194mm　1/16	**印　　张**：5.5
版　　次：2017 年 9 月第 1 版	**印　　次**：2017 年 9 月第 1 次印刷
字　　数：150 千字	**定　　价**：58.00 元

ISBN 978-7-5130-5093-7

编审委员会

统　　　筹：国家知识产权局专利局初审及流程管理部青年工作组

统筹组成员：吴登侣　毛晓鹏

编撰组成员：（按姓氏笔画排列）

丁　玮　丁子剑　于佳晶　毛晓鹏　卢　佳　田　明　付　强　师严涛　刘　佳

刘文雯　刘时娇　李炜倩　杨媛媛　吴登侣　余梅霜　张　洁　张其文　陈仰平

庞　琳　郎　乐　胡　扬　勇　飞　桂　林　徐涪浩　康　飞　康德地　章洁桦

梁　爽　梁玉琴　羡晨静　鄢　波　潘　晓　冀　梦　魏　欢

审稿组成员：（按姓氏笔画排列）

王星跃　王靖梅　石　莉　申江华　刘　巍　刘丽君　苏春波　李　享　李　莉

杨　兴　张　颖　张莉丽　林　梅　赵　霞　胡泽建　姚晓红　郭　强　唐　凯

序

这是一本很有意思的专利审查业务图书。

说它有意思，首先是因为它的展现形式。在知识产权学术领域，以文字为主要展示形式的学术专著司空见惯，而纯粹以图片形式展示业务内容的作品却很少见。其次，以简洁、直观的图片解析专利审查流程中的业务，体现出编者为广大创新主体提供便捷、有效的专利审查流程服务的良苦用心。

2008年，《国家知识产权战略纲要》发布实施，为我国知识产权事业的发展铺设了新的蓝图；2015年12月，《国务院关于新形势下加快知识产权强国建设的若干意见》(以下简称《意见》)发布，预示着在未来一段时间内，知识产权强国建设将成为所有知识产权从业人员共同的奋斗目标。《意见》中明确指出："优化专利和商标的审查流程和方式，实现知识产权在线登记、电子申请和无纸化审批。"这是专利审查流程优化的政策要求。

从现实情况来看，当前我国的自主创新能力不断提升，而对于广大创新主体来说，专利审查流程存在"点多、线长、面广"的特点，业务内容相对繁杂。广大创新主体对专利审查流程了解有限，无法通过阅读《专利审查指南》等规范性文件迅速掌握具体业务的办理流程。因此在专利申请及专利维权的过程中会遇到"摸不着路"的情况，甚至会因为不熟悉相关程序导致权利丧失。

在这样的背景下，自2015年起，国家知识产权局专利局初审及流程管理部围绕国家知识产权局党组的要求和部署，在不断优化专利审查流程的同时，结合广大创新主体的实际业务需求，组织部门青年业务骨干认真梳理现有专利审查流程，分批次绘制中英文两个版本共76幅"一张图看懂专利审批流程"系列图，内容基本实现专利审查流程的全覆盖。其中小部分图在相关微信公众号上发布后，引发社会热烈反响，短短3天转发阅读量突破5万人次。集结成书的需求，日渐迫切。于是《一张图看懂专利审批流程》顺势而出。

本书首先向读者概要展示专利申请审查流程，再按照流程顺序，依次展示受理阶段、发明专利初步审查阶段、PCT申请国际阶段及国家阶段、办理登记手续和专利权维持阶段的主要手续，以及相关的流程服务和行政复议程序。

本书仅供社会各界人士快速了解相关专利审查流程，具体审查要求，仍以相关法律法规为准。欢迎社会各界人士提出宝贵意见和建议。我们将根据社会需求和专利审查制度的变化，继续完善该系列图片，为大众创业、万众创新提供持续、有力的支持。

2017年2月

目 录

CONTENTS

导　语

现在让我们共同开启愉快的专利审批流程之旅吧！

专利申请审批流程

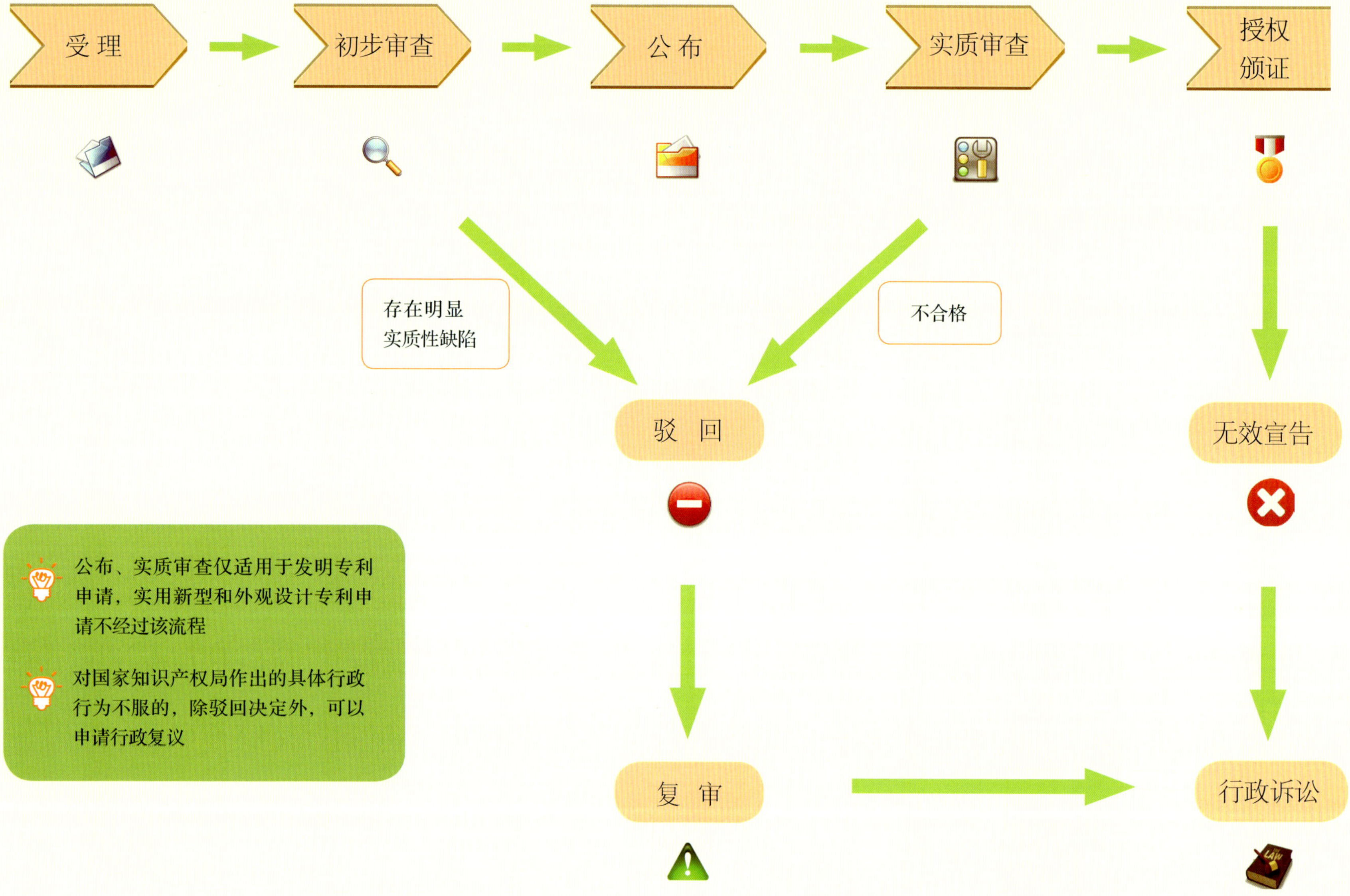

受 理 → 初步审查 → 公 布 → 实质审查 → 授权颁证

存在明显
实质性缺陷

不合格

驳 回

无效宣告

公布、实质审查仅适用于发明专利申请，实用新型和外观设计专利申请不经过该流程

对国家知识产权局作出的具体行政行为不服的，除驳回决定外，可以申请行政复议

复 审 → 行政诉讼

Patent Examining Procedure

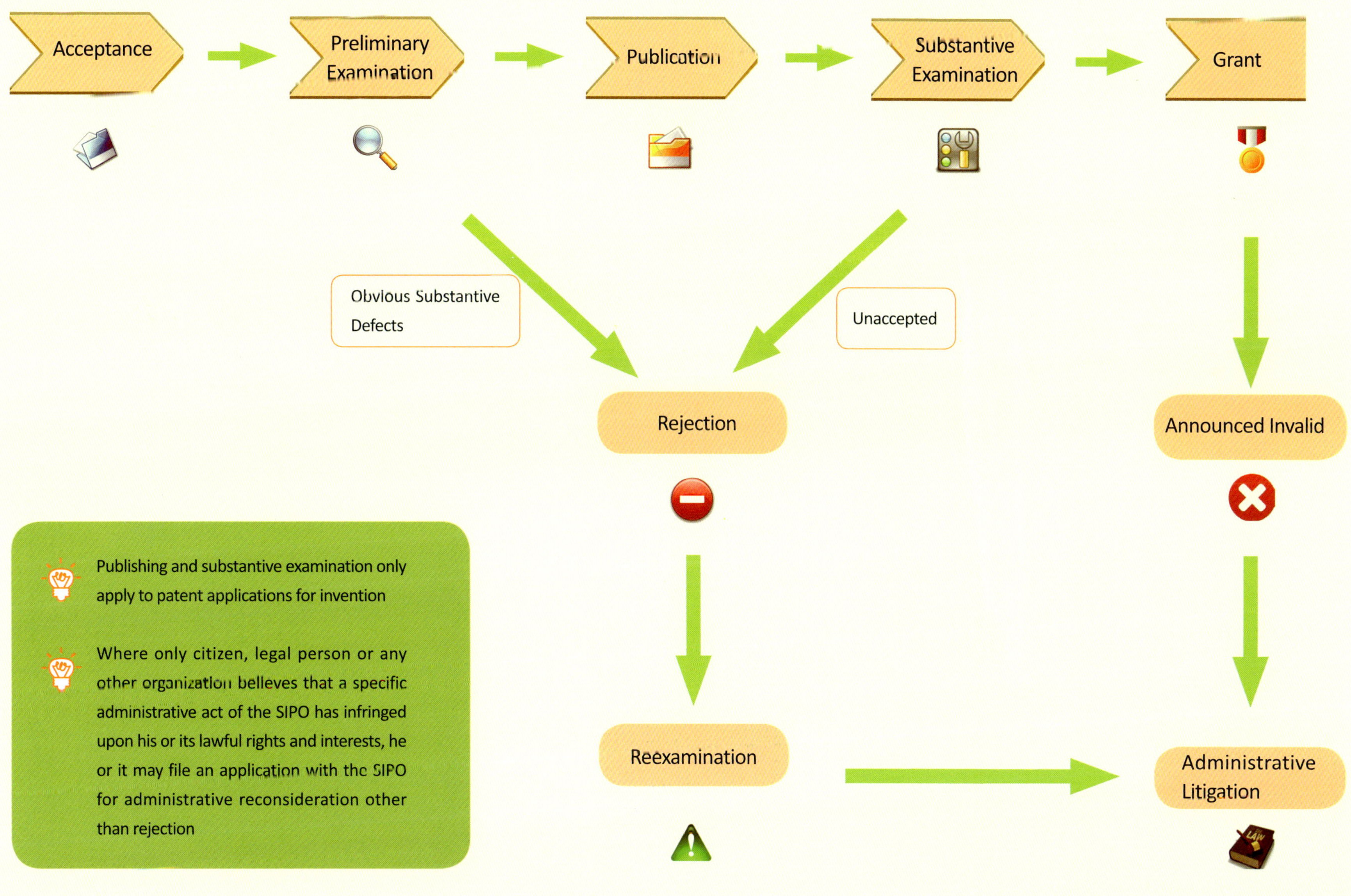

Acceptance → Preliminary Examination → Publication → Substantive Examination → Grant

Obvious Substantive Defects

Unaccepted

Rejection

Announced Invalid

Publishing and substantive examination only apply to patent applications for invention

Where only citizen, legal person or any other organization believes that a specific administrative act of the SIPO has infringed upon his or its lawful rights and interests, he or it may file an application with the SIPO for administrative reconsideration other than rejection

Reexamination

Administrative Litigation

专利申请受理

申请文件 提交方式 ➥ 纸件提交

（1）当面提交（专利局及其代办处）
（2）邮寄提交

＋  电子提交

必要的 申请文件 ➥

（1）发　明：请求书、权利要求书、说明书

（2）实用新型：请求书、权利要求书、说明书、说明书附图

（3）外观设计：请求书、图片或照片、简要说明

不受理的 情形举例 ➥

 （1）申请文件未使用中文

 （2）手工书写，字迹或线条 模糊，有涂改，易擦除

 （3）直接从境外邮寄的文件

 （4）专利申请类别不明确， 分案申请改变原申请类别

Patent Filing

 **Manner of Filing**

Paper

(1) Directly submit to SIPO or local agencies

(2) Mailing

 File online

 **Required Documents**

(1) Invention: request, claims and description

(2) Utility Model: request, claims, description and drawings

(3) Design: request, drawings or photographs and a brief description

 **Unaccepted**

(1) Not in Chinese

(2) Handwritten ,vague, altered, or easy to erase

(3) Documents delivered overseas

(4) Patent type is not clear, divisional application requesting different type from origin

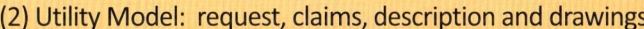
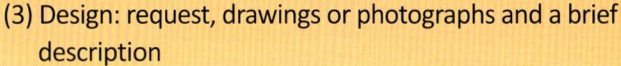

专利电子申请用户注册

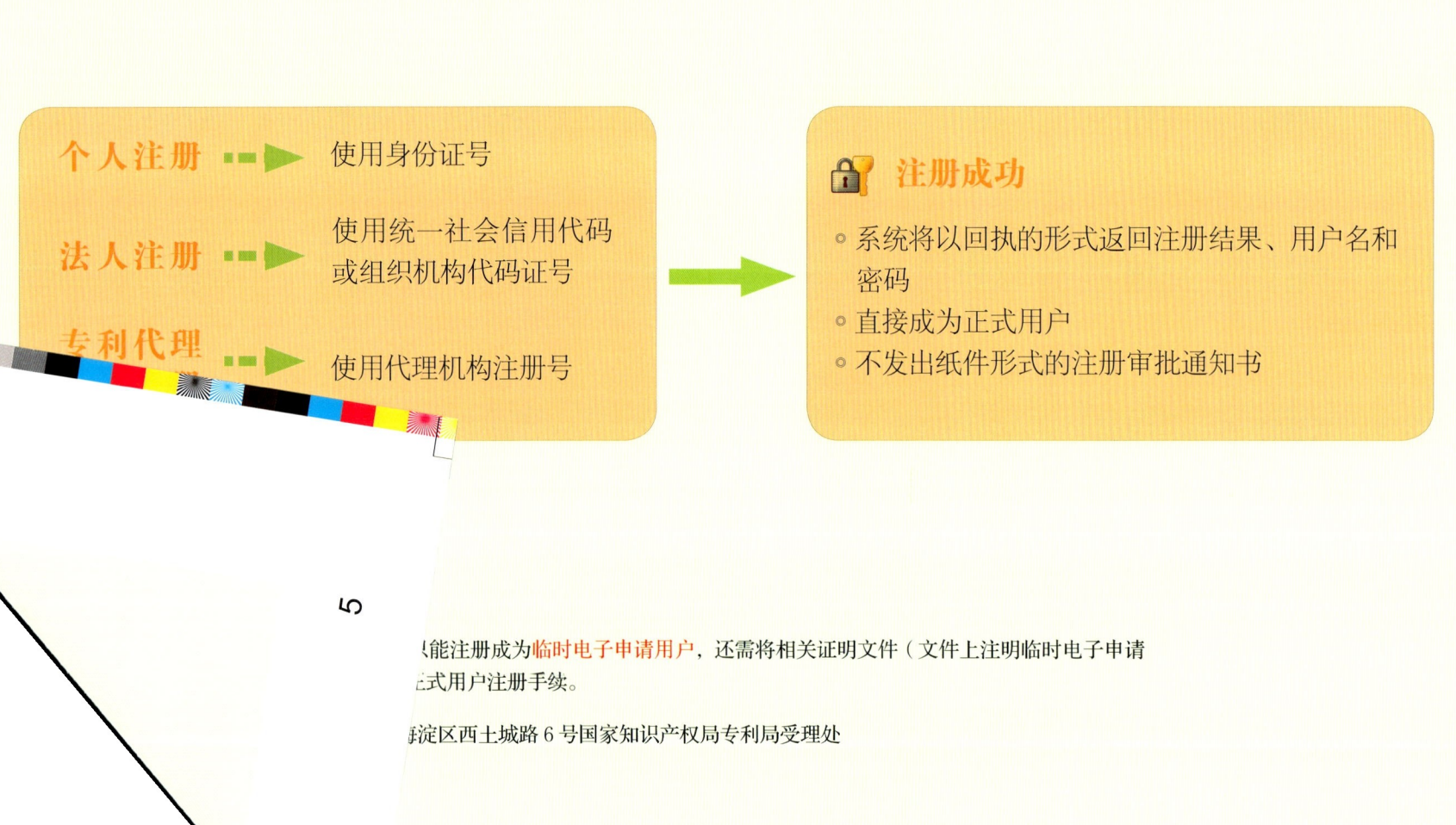

在中国专利电子申请网
办理注册手续 ➡ www.cponline.gov.cn

个人注册 ➡ 使用身份证号

法人注册 ➡ 使用统一社会信用代码
或组织机构代码证号

专利代理 ➡ 使用代理机构注册号

🔒 注册成功

◦ 系统将以回执的形式返回注册结果、用户名和密码
◦ 直接成为正式用户
◦ 不发出纸件形式的注册审批通知书

5

……只能注册成为临时电子申请用户，还需将相关证明文件（文件上注明临时电子申请……

……正式用户注册手续。

……海淀区西土城路 6 号国家知识产权局专利局受理处

User Registration of Electronic Application

◢ **Visit China Patent Electronic Application Website** ▶ www.cponline.gov.cn

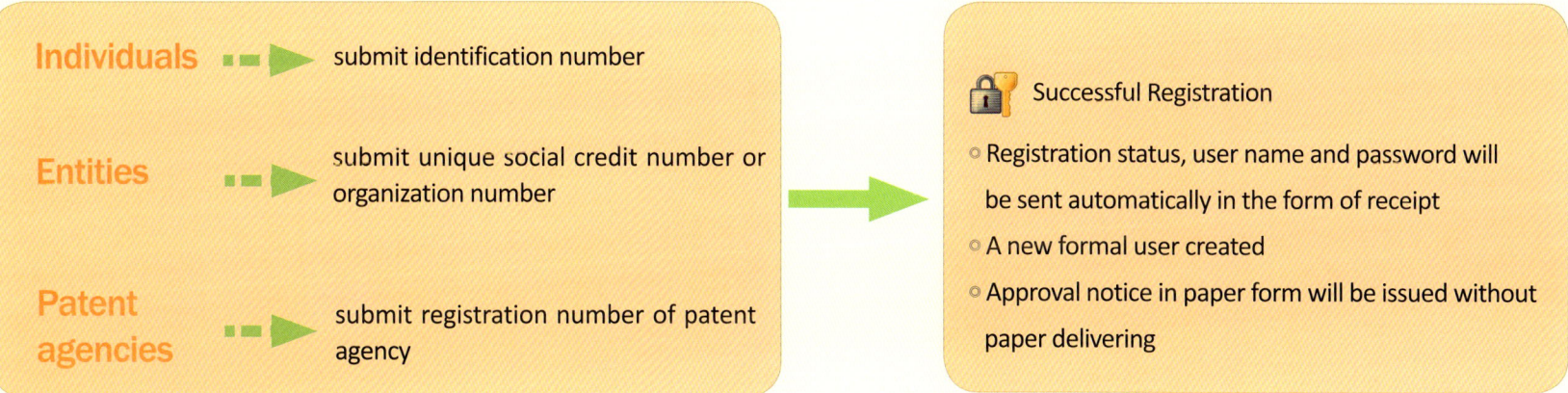

Individuals ▶ submit identification number

Entities ▶ submit unique social credit number or organization number

Patent agencies ▶ submit registration number of patent agency

🔓 Successful Registration

○ Registration status, user name and password will be sent automatically in the form of receipt

○ A new formal user created

○ Approval notice in paper form will be issued without paper delivering

 If using other certification number, temporary electronic application users will be registered, but relevant documents (temporary electronic application user account shall be indicated) are needed to mail to the patent office for formal registration procedures

 Mailing address: Acceptance Division, State Intellectual Property Office, No. 6, Xitucheng Road, Haidian District, Beijing

Postcode : 100088

7

电子申请在线业务办理平台

◾ 平台能办理哪些业务

- 在线提交新申请文件，接收、答复通知书
- 办理著录项目变更、补正、恢复权利请求、延长通知书期限、撤回优先权、陈述意见、缴纳费用等手续
- 账号和证书管理、案件管理、在线证明文件备案……

💡 DAS 和 PPH 业务需要通过 CPC 客户端提交或者基于交互式平台提交纸件请求

◾ 从哪里登录平台

www.cponline.gov.cn

◾ 平台有哪些新亮点

- 在线自助注册电子申请用户
- 实时提供直观提示信息
- 文件提交前自动纠错
- 部分手续即时审批
- 批量办理著录项目变更手续
- 证明文件在线备案
- 支持多层级用户管理

◾ 在线与离线电子提交形式可以互相转换吗

在电子申请客户端提交离线专利电子申请并被受理后，可以转为在线专利电子申请；除另有规定外，在线电子申请不能转为离线电子申请，不能以离线电子形式提交文件

Online Patent Application Platform

Provided Services

- Submission of patent application documents, receiving and replying to the patent notifications
- Submission of request for changes in bibliographic data, rectification, request for restoration of right, extension of time limit, withdrawal of claim to the right of priority, statement of observations, and payment of patent fees, etc.
- Management of accounts and e-certificates, management of patent cases, record of documents, etc.

Formalities regarding DAS and PPH shall be gone through by CPC or by interactive platform in paper form

Access to Platform

www.cponline.gov.cn

Several Strengths

- Allow self-registration of electronic application user
- Visual message will be provided timely
- Allow automatic error correction before the submission of documents
- Provide real-time examination on several formalities
- Allow batch processing in bibliographic data changes
- Allow online record
- Allow management of multi-level users

Transformation between Online and Offline Application Form

When an offline patent application is submitted and accepted, it will be transformed to online form upon request and approval. Unless otherwise specified, online application will not be transformed to offline form, and will not submit any documents offline

9

使用电子申请客户端收发文件

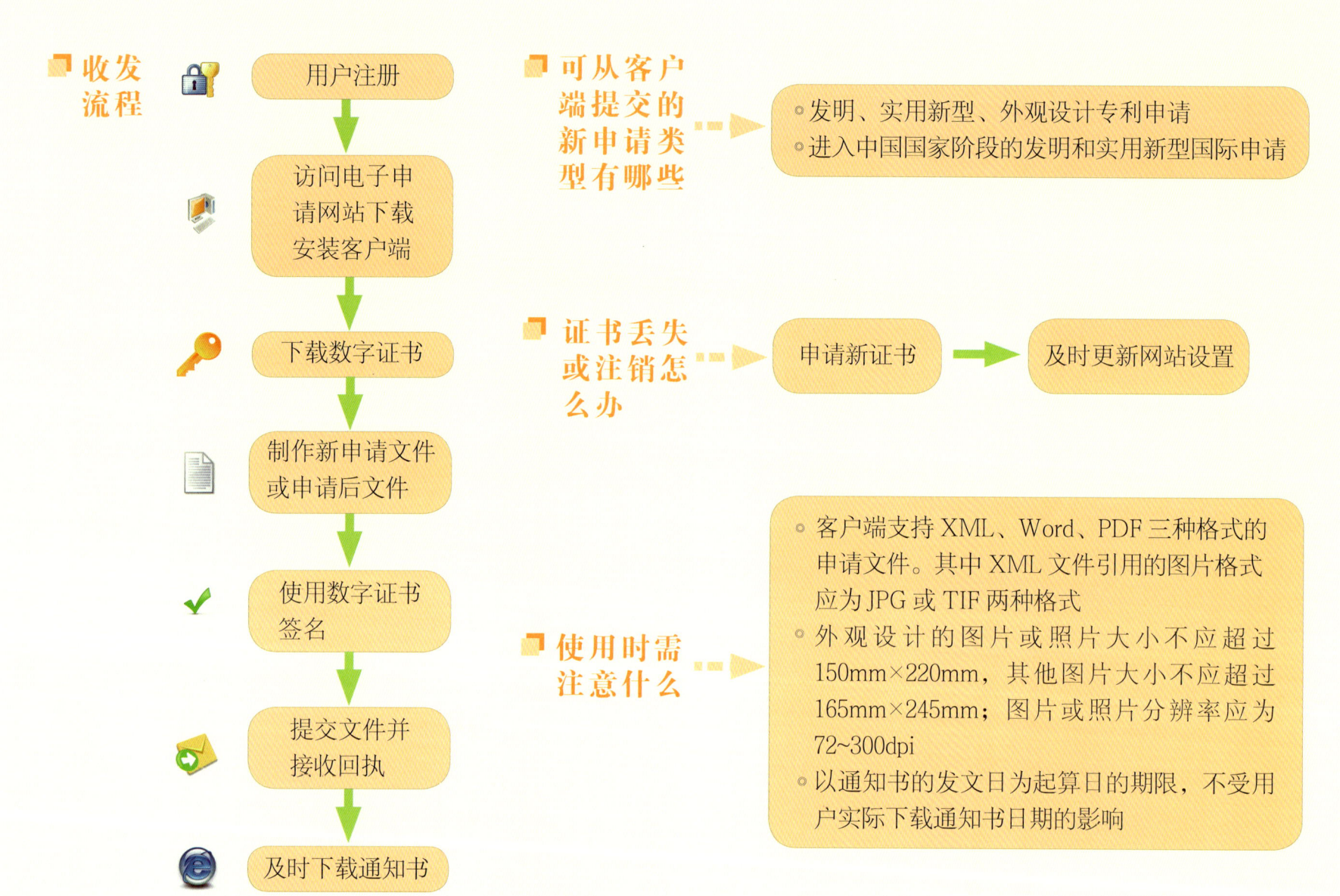

收发流程

用户注册

↓

访问电子申请网站下载安装客户端

↓

下载数字证书

↓

制作新申请文件或申请后文件

↓

使用数字证书签名

↓

提交文件并接收回执

↓

及时下载通知书

可从客户端提交的新申请类型有哪些

- 发明、实用新型、外观设计专利申请
- 进入中国国家阶段的发明和实用新型国际申请

证书丢失或注销怎么办

申请新证书 → 及时更新网站设置

使用时需注意什么

- 客户端支持 XML、Word、PDF 三种格式的申请文件。其中 XML 文件引用的图片格式应为 JPG 或 TIF 两种格式
- 外观设计的图片或照片大小不应超过 150mm×220mm，其他图片大小不应超过 165mm×245mm；图片或照片分辨率应为 72~300dpi
- 以通知书的发文日为起算日的期限，不受用户实际下载通知书日期的影响

Transceiver of Document Transmission on Electronic Application Platform

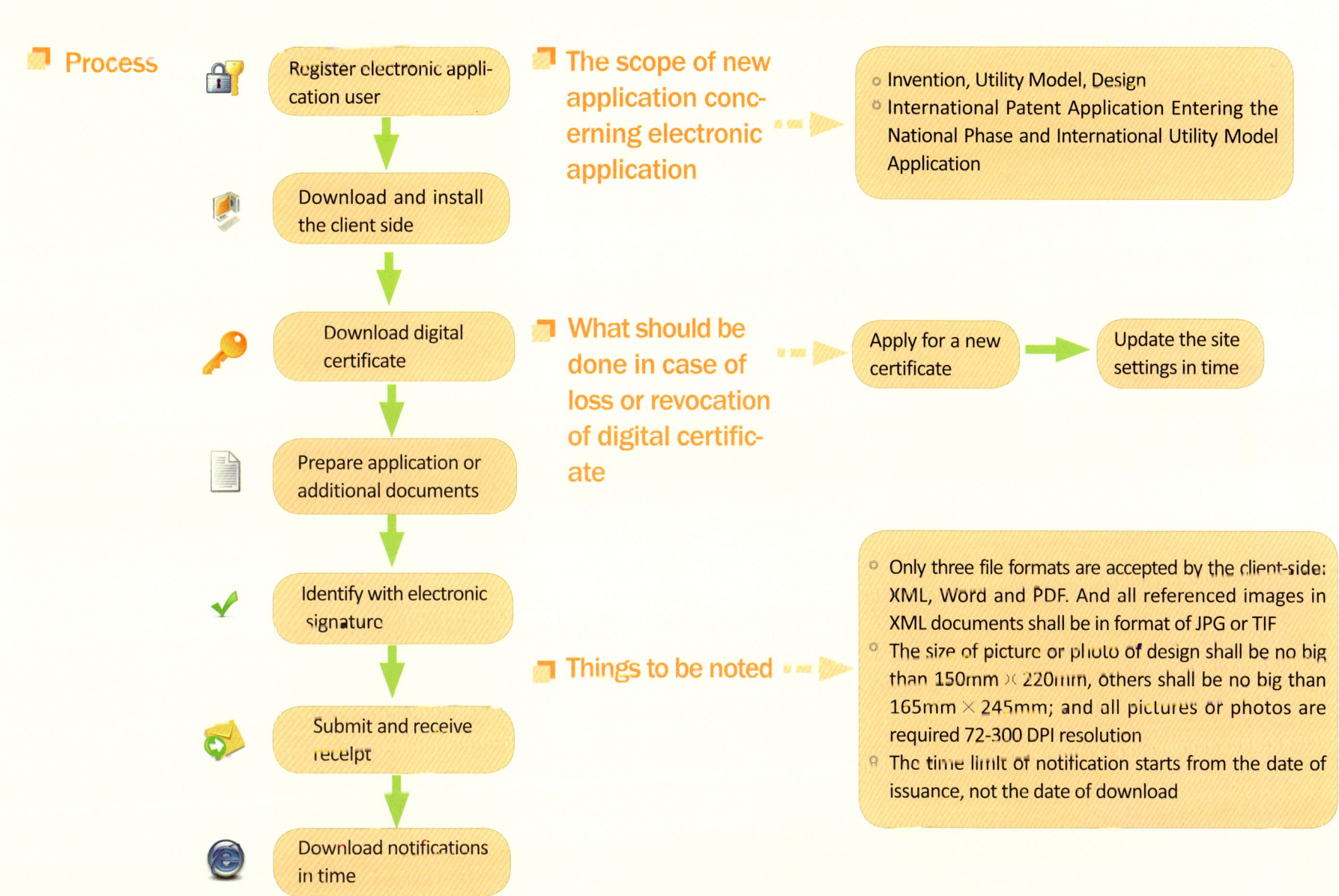

Process

- Register electronic application user
- Download and install the client side
- Download digital certificate
- Prepare application or additional documents
- Identify with electronic signature
- Submit and receive receipt
- Download notifications in time

The scope of new application concerning electronic application
- Invention, Utility Model, Design
- International Patent Application Entering the National Phase and International Utility Model Application

What should be done in case of loss or revocation of digital certificate
- Apply for a new certificate → Update the site settings in time

Things to be noted
- Only three file formats are accepted by the client-side: XML, Word and PDF. And all referenced images in XML documents shall be in format of JPG or TIF
- The size of picture or photo of design shall be no big than 150mm × 220mm, others shall be no big than 165mm × 245mm; and all pictures or photos are required 72-300 DPI resolution
- The time limit of notification starts from the date of issuance, not the date of download

在客户端提交申请后文件的权限人

▪ 谁是提交权限人

情　况	权限人
一个申请人且未委托专利代理机构	申请人
多个申请人且未委托专利代理机构	代表人
一个或多个申请人委托专利代理机构	专利代理机构

▪ 提交权限的意义

需要申请人、代表人或专利代理机构办理的审查手续，电子申请系统仅允许提交权限人提交该手续相关文件，非提交权限人提交的手续文件在提交时拒收

▪ 不限制提交权限的情形

- 恢复权利请求
- 退款请求
- 中止请求
- 社会公众提出的意见陈述相关文件
- 专利权无效宣告请求
- 专利权评价报告请求
- 实用新型检索报告请求

▪ 这些情况会导致提交权限人变更

- 申请人或专利权人变更
- 委托或更换代理机构
- 解除或辞去委托
- 视为未委托代理机构
- 未委托代理机构且代表人变更

The Submission Authority of Additional Documents

Who have the authority of submission

Scenario	Authorizer
Single applicant without appointing a patent agency	The applicant
Two or more applicants without appointing a patent agency	The representative
One or more applicants with a patent agency appointed	The appointed agency

The meaning of submission authority control

Only the authorizer is permitted to submit relative documents regarding the formalities that need to be handled by applicants or patent agencies. Documents submitted by non-authorized party will be rejected when submission

Formalities with unrestricted submission permission

- Request for restoration of right
- Request for Refund
- Request for Suspension
- Statement of opinion submitted by the public
- Request for Invalidation
- Request for Evaluation Report of Patent
- Request for Search Report of Utility Model Patent

Circumstances of changes in authorizer

- Change of applicant or patentee
- Appointment or change of patent agency
- Dissolution or resignation of appointment
- Patent agency deemed not to have been appointed
- Change in the representative and no patent agency appointed

13

专利费用减缴

 **谁能请求费用减缴** ┈▶

（1）上年度月均收入低于3500元（年4.2万元）的个人

（2）上年度企业应纳税所得额低于30万元的企业

（3）事业单位、社会团体、非营利性科研机构

哪些费种可以减缴

- 申请费
- 发明专利申请实质审查费
- 复审费
- 授权当年起6年内年费

证明文件如何备案 ┈▶

完成备案后，将以下证明文件提交到指定代办处：

通过专利事务服务系统提交减缴请求并经审批备案，一个自然年度内再次请求减缴，仅需提交减缴请求，无须提交证明文件

（1）所在单位出具的年度收入证明，无固定工作的提交县级民政部门或乡镇人民政府（街道办事处）出具的经济困难情况证明

（2）所在单位出具的年度收入证明，无固定工作的提交县级民政部门或乡镇人民政府（街道办事处）出具的经济困难情况证明

（3）提交法人证明材料复印件

享受多大减缴比例

- 一个申请人：减缴85%
- 两个及以上申请人：减缴70%

如何请求费用减缴 ┈▶

新申请：在请求书中勾选"请求费减且已完成费减资格备案"，并写明证件号码（即备案号）
申请后：提交费用减缴请求书，并写明证件号码（即备案号）减缴申请费的请求应当与专利申请同时提出，其他费用减缴请求可在缴费期限届满两个半月之前提出

Reduction of Patent Fees

◪ Who can request

→ Individuals, whose average monthly salary is no more than 3500 CNY (or annual income is no more than 42000 CNY) in the last calendar year

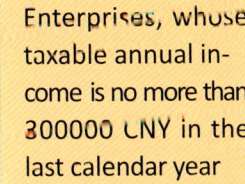

Enterprises, whose taxable annual income is no more than 300000 CNY in the last calendar year

Public institutions, Social Groups and non-profit Public Research organizations

◪ Fees can be reduced

○ Filing fees
○ Fee for initiating substantive examination
○ Fee for re-examination
○ Annuities (for six consecutive years since the grant of a Chinese patent including the one due in the registration fees)

◪ How to record qualifications for the fee reduction

→ When the recordal is done, below documents are required to be delivered to the local patent agencies:

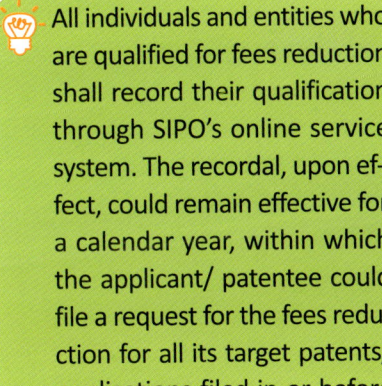

All individuals and entities who are qualified for fees reduction shall record their qualification through SIPO's online service system. The recordal, upon effect, could remain effective for a calendar year, within which the applicant/ patentee could file a request for the fees reduction for all its target patents/ applications filed in or before this year case by case

(1) Income certificate issued by his or her organization. If the person does not have permanent job, the attestation of financial difficulties issued by local civil administration department is required

(2) Notarized copy of the Income Tax on enterprises annual tax return of last year; during the period of final settlement, the income tax on enterprises annual tax return of the year before last year is required

(3) Notarized copy of the certificate attesting legal status

◪ Reduction ratio

○ One single applicant:Reduced by 85%
○ Two or more applicants:Reduced by 70%

◪ How to request on case-by-case basis

→ New patent application: Check "Qualifications have been recorded, hereby request for fee reduction" in the patent request

After application date: submit request of reduction of payment of patent fees, quoting the recordal number. The reduction of filing fees shall be requested simultaneously with application, and reduction of other official fees shall be requested before two and half months from the payment deadline

专利缴费

当面缴纳 ...▶

地 点	+	方 式
专利局及代办处窗口		现金、支票、刷卡

网上缴费 ...▶ 电子申请用户登录 www.cponline.gov.cn，使用网上缴费系统

银行汇付 ...▶
开户银行：中信银行北京知春路支行
户 名：中华人民共和国国家知识产权局专利局
账 号：7111710182600166032

> 💡 银行实际汇出日为缴费日

> 💡 汇款单附言栏中写明正确的
> 申请号及费用名称
> 💡 补充缴费信息的，以专利局
> 收到正确信息之日为缴费日

邮局汇付 ...▶
收款人姓名：国家知识产权局专利局收费处
地址（邮编）：北京市海淀区西土城路 6 号（100088）
传 真：010-62084312

> 💡 邮局汇出的邮戳日为缴费日

Payment

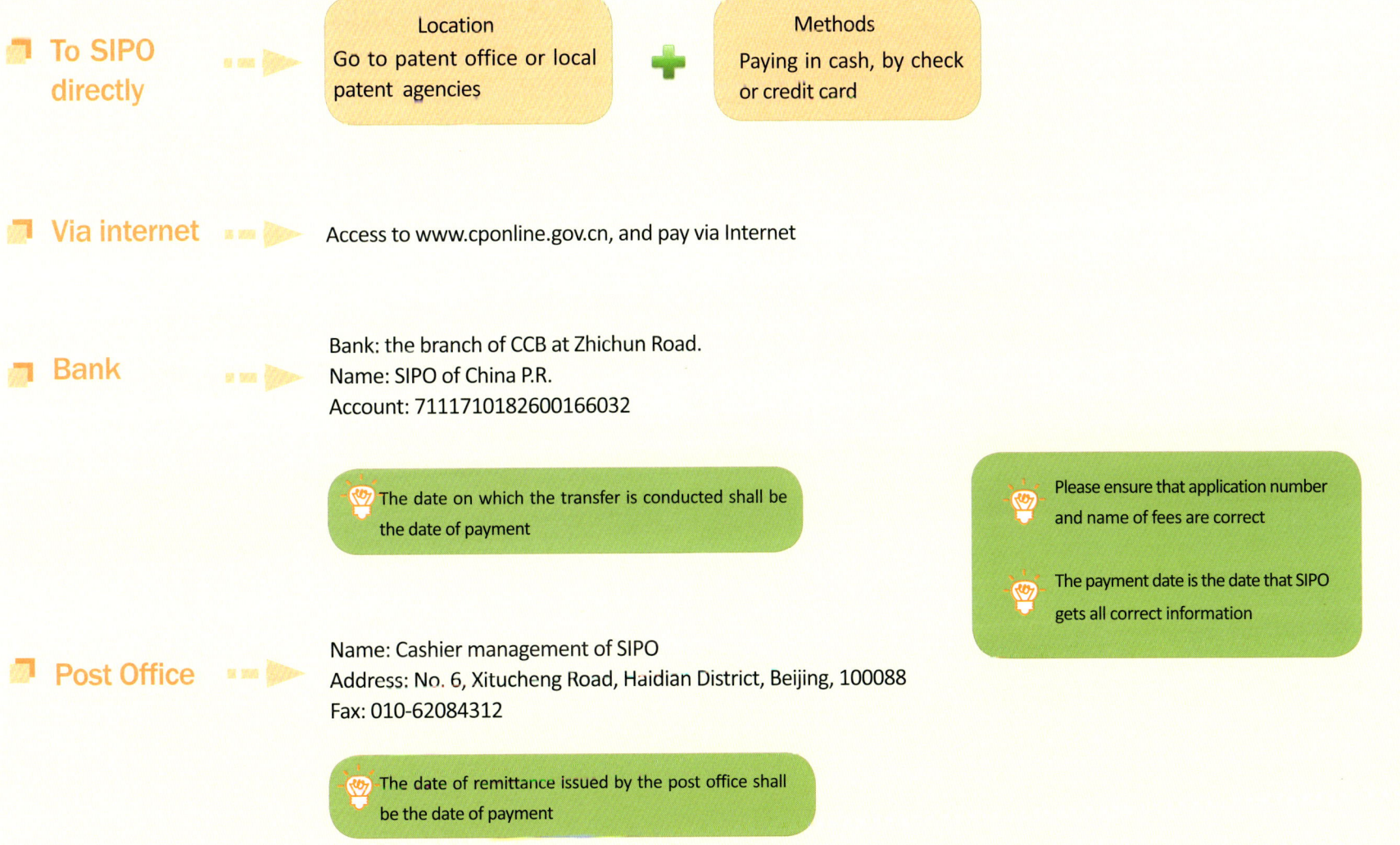

To SIPO directly

Location	Methods
Go to patent office or local patent agencies	Paying in cash, by check or credit card

Via internet

Access to www.cponline.gov.cn, and pay via Internet

Bank

Bank: the branch of CCB at Zhichun Road.
Name: SIPO of China P.R.
Account: 71117101826000166032

The date on which the transfer is conducted shall be the date of payment

Please ensure that application number and name of fees are correct

The payment date is the date that SIPO gets all correct information

Post Office

Name: Cashier management of SIPO
Address: No. 6, Xitucheng Road, Haidian District, Beijing, 100088
Fax: 010-62084312

The date of remittance issued by the post office shall be the date of payment

发明专利初步审查

发明初审 审什么

申请文件及其他 文件形式审查 ➕ 明显实质性 缺陷审查 ➕ 有关费用 审查 ➕ 有关期限审查

收到通知 怎么做

办理手续补正通知书 补正通知书或审查意见通知书

提交补正书 + 补正文件 提交补正书 / 意见陈述书 + 补正文件

可能的结 果有哪些

逾期不答复 ➡ 手续视为未提出 ➡ 专利申请视为撤回

缺陷不能克服 ➡ 手续视为未提出 ➡ 驳回

缺陷已经克服 ➡ 手续合格 ➡ 初审合格

Preliminary Examination of Invention

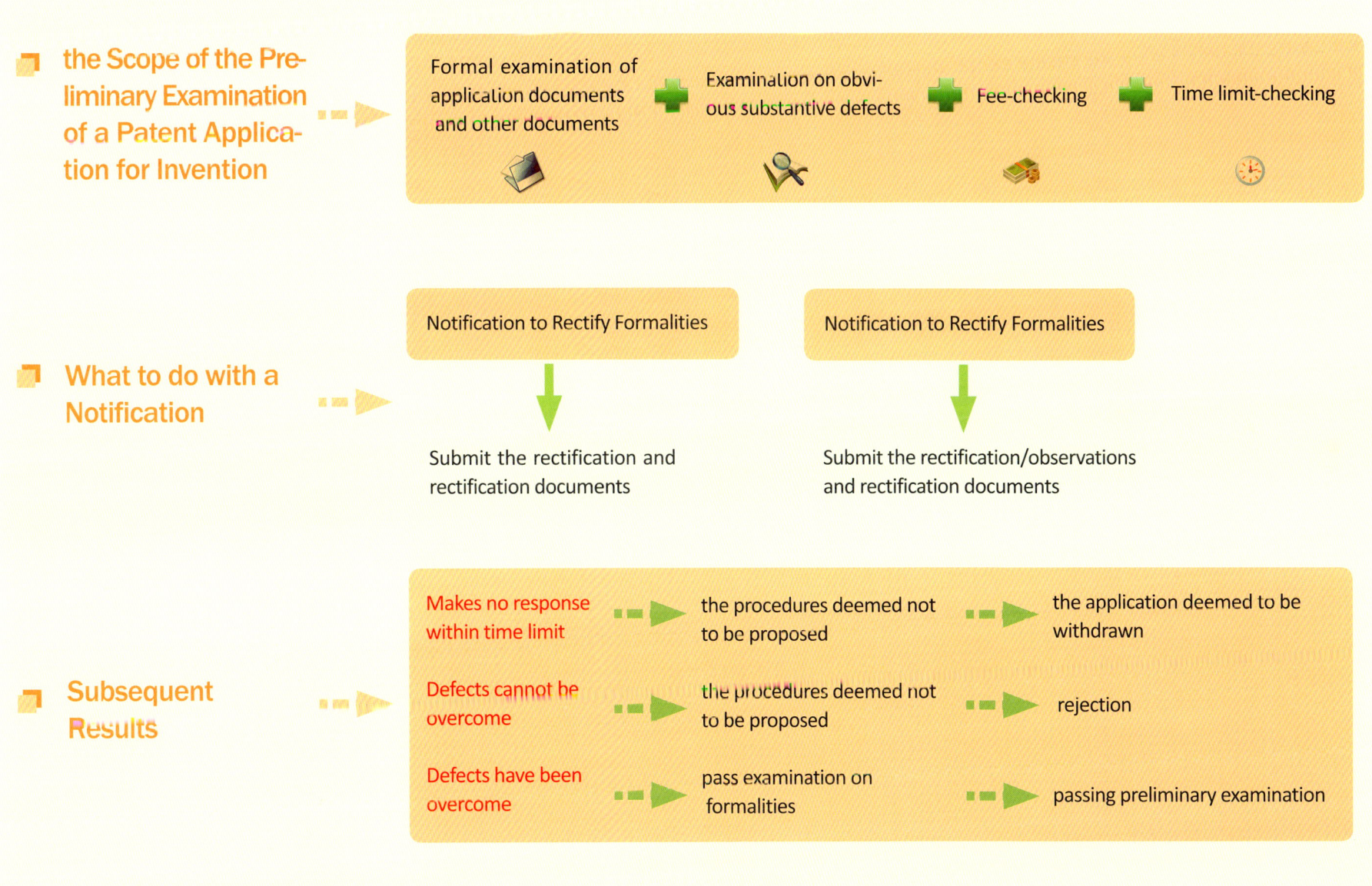

the Scope of the Preliminary Examination of a Patent Application for Invention

Formal examination of application documents and other documents

Examination on obvious substantive defects

Fee-checking

Time limit-checking

What to do with a Notification

Notification to Rectify Formalities

Submit the rectification and rectification documents

Notification to Rectify Formalities

Submit the rectification/observations and rectification documents

Subsequent Results

Makes no response within time limit → the procedures deemed not to be proposed → the application deemed to be withdrawn

Defects cannot be overcome → the procedures deemed not to be proposed → rejection

Defects have been overcome → pass examination on formalities → passing preliminary examination

电子交换获取优先权副本

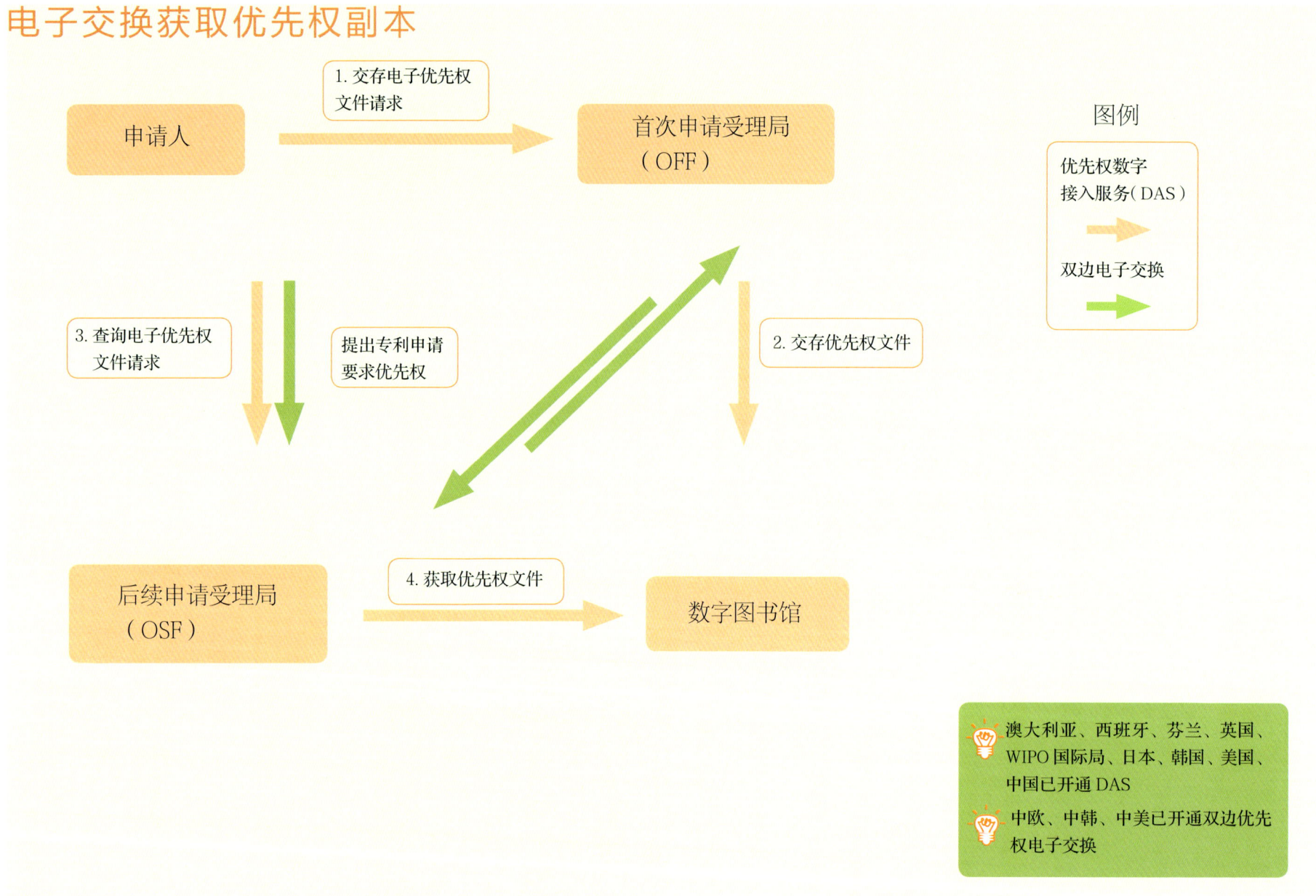

申请人

1. 交存电子优先权文件请求

首次申请受理局（OFF）

3. 查询电子优先权文件请求

提出专利申请要求优先权

2. 交存优先权文件

后续申请受理局（OSF）

4. 获取优先权文件

数字图书馆

图例

优先权数字接入服务（DAS）

双边电子交换

澳大利亚、西班牙、芬兰、英国、WIPO 国际局、日本、韩国、美国、中国已开通 DAS

中欧、中韩、中美已开通双边优先权电子交换

E-Exchange Priority Copy

Applicant

1.Request to storage priority documents

Office of First Filing(OFF)

3.Search the request of priority

Filing a patent application and claiming the right of priority

2.Deliver priority documents

Office of Second Filing(OSF)

4.Obtain the priority documents

Digital Access Service (DAS)

Figure

Digital Access Service (DAS)

Bilateral Electronic Exchange (BEE)

Australia, Spain, Finland, UK, WIPO, Japan, Korea, USA, China can provide DAS service

China and EU, China and Korea, China and USA have signed a bilateral agreement to exchange priority copy electronically

要求外国优先权的手续

相同主题在外国的首次申请 ——[发明或者实用新型]—— 12个月内在中国提出发明或实用新型专利申请

相同主题在外国的首次申请 ——[外观设计]—— 6个月内在中国提出外观设计专利申请

■ **外国优先权的作用**

> 方便《巴黎公约》成员国国民或居民在不同成员国获得专利权；在后申请在是否具备新颖性、创造性等方面被视为是在首次申请的申请日提出；在后申请公布、提出实质审查期限等方面被视为自优先权日起算

■ **要求外国优先权声明**

（1）声明的时间：提出专利申请时

（3）优先权费用：每项80元

（2）声明的内容：
- 在先申请的申请日
- 在先申请的申请号
- 原受理机构名称

■ **需要提交哪些文件**

> 在先申请文件副本：
> 提交时间：在后申请日起3个月内
> 提交方式：纸件方式、电子方式、电子交换（DAS、双边）
> 已提交过，需要再次提交的，在中文题录中注明原件所在案卷的申请号

> 在先申请文件副本中文题录内容：
> - 原受理机构名称
> - 在先申请的申请日
> - 在先申请的申请号
> - 在先申请的申请人

> 特殊情况：
> 当在后申请的申请人不是在先申请的申请人全体或部分成员时，需于在后申请日起3个月内，提交优先权转让证明、优先权转让证明中文题录

The Procedure of Claiming Foreign Priority

| The first application of the same subject matter filing in foreign country | Inventions or utility | File an application for invention or utility model in China within 12 months | The first application of the same subject matter filing in foreign country | Designs | File an application for design patent in China within 6 months |

 **Role of foreign priority**

> National or resident of the Paris Convention could obtain patent right in different member states; In examination of novelty and inventive step etc, the subsequent application shall be deemed to be the made on the priority date ;The time limit for publication and substantive examination request of subsequent application will be started from the priority date

 **Declaration of claiming foreign priority**

(1) Time to make a declaration:
   when the patent application is filed

(3) Fee for Claiming Priority :
    80 CNY per item

(2) The contents of declaration:
   o Filing date of the previous application
   o Application number of the previous application
   o Name of the original authority

Documents need to be submitted

> Copy of previous application documents:
> Submit time: within 3 months from the filing date of the subsequent application
> Submit ways: paper form, electronic form, electronic transmission
> The copy required to be submitted again, Indicates the application number of the subsequent application of which the certified copy of the previous application document is deposited

> the Chinese translation of the extract content:
> o Name of the original authority
> o Filing date of the previous application
> o Application number of the previous application
> o Applicant of the previous application

> Special Case:
> If the applicant(s) of the subsequent application shall be not the same as the applicant(s) ,or at least one of the previous applicant(s),it is required to submit a document certifying the assignment of the right of priority and the Chinese translation of the extract of above within 3 months from the filing date of the subsequent application

要求本国优先权的手续

- **本国优先权** ··▶ 相同主题在中国的首次申请 ——▶ 12 个月内再次提出发明或实用新型专利申请并要求优先权

> 在先申请自在后申请提出之日起即视为撤回

- **本国优先权的作用** ··▶ 实现发明和实用新型专利申请类型的转换；补充完善首次申请或补救在先申请审查程序中的失误；在后申请在是否具备新颖性、创造性等方面被视为是在首次申请的申请日提出，在后申请在公布期限、提出实质审查期限等方面被视为自优先权日起算。

- **在先申请应满足什么条件** ··▶
  - （1）发明或者实用新型
  - （2）尚未授权
  - （3）不是分案申请
  - （4）没有享有优先权

- **要求本国优先权声明** ··▶
  - （1）声明的时间：提出专利申请时
  - （2）声明的内容：
    - ◦在先申请的申请日
    - ◦在先申请的申请号
    - ◦原受理机构名称
  - （3）优先权费用：每项 80 元

- **在后申请的申请人** ··▶
  - （1）应当与在先申请的申请人完全一致
  - （2）不一致的应提交：优先权转让证明
  - （3）提交时间：在后申请日起 3 个月内

The Procedure of Claiming Domestic Priority

Domestic Priority ▸ The first application of the same subject matter filing in China 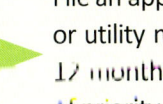 File an application for invention or utility model in China within 12 months and claim the right of priority

 The previous application shall be deemed to have been withdrawn from the date on which the subsequent application is filed

Role of domestic priority ▸ To realize the conversion between invention and utility model; To complete and perfect the first application; In examination of novelty and inventive step etc., the subsequent application shall be deemed to be made on the priority date ;The time limit for publication and substantive examination request of the subsequent application will be calculated from the priority date

The previous application should ▸
(1) Be invention or utility model (2) Have not yet been granted for patent right

(3) Not be a divisional application (4) Not enjoy any priority

Declaration of claiming domestic priority ▸
(1) Time to make a declaration:
  when the patent application is filed
(3) Fee for Claiming priority :
  80 CNY per item

(2) The contents of declaration:
  ○ Filing date of the previous application
  ○ Application number of the previous application
  ○ Name of the authority with which the application was first filed

The applicant(s) of the subsequent application ▸
(1) Shall be the same as that of the previous application
(2) Otherwise shall submit : a document certifying the assignment of the right of priority
(3) Time of submission: within 3 months from the filing date of the subsequent application

分案申请递交时间

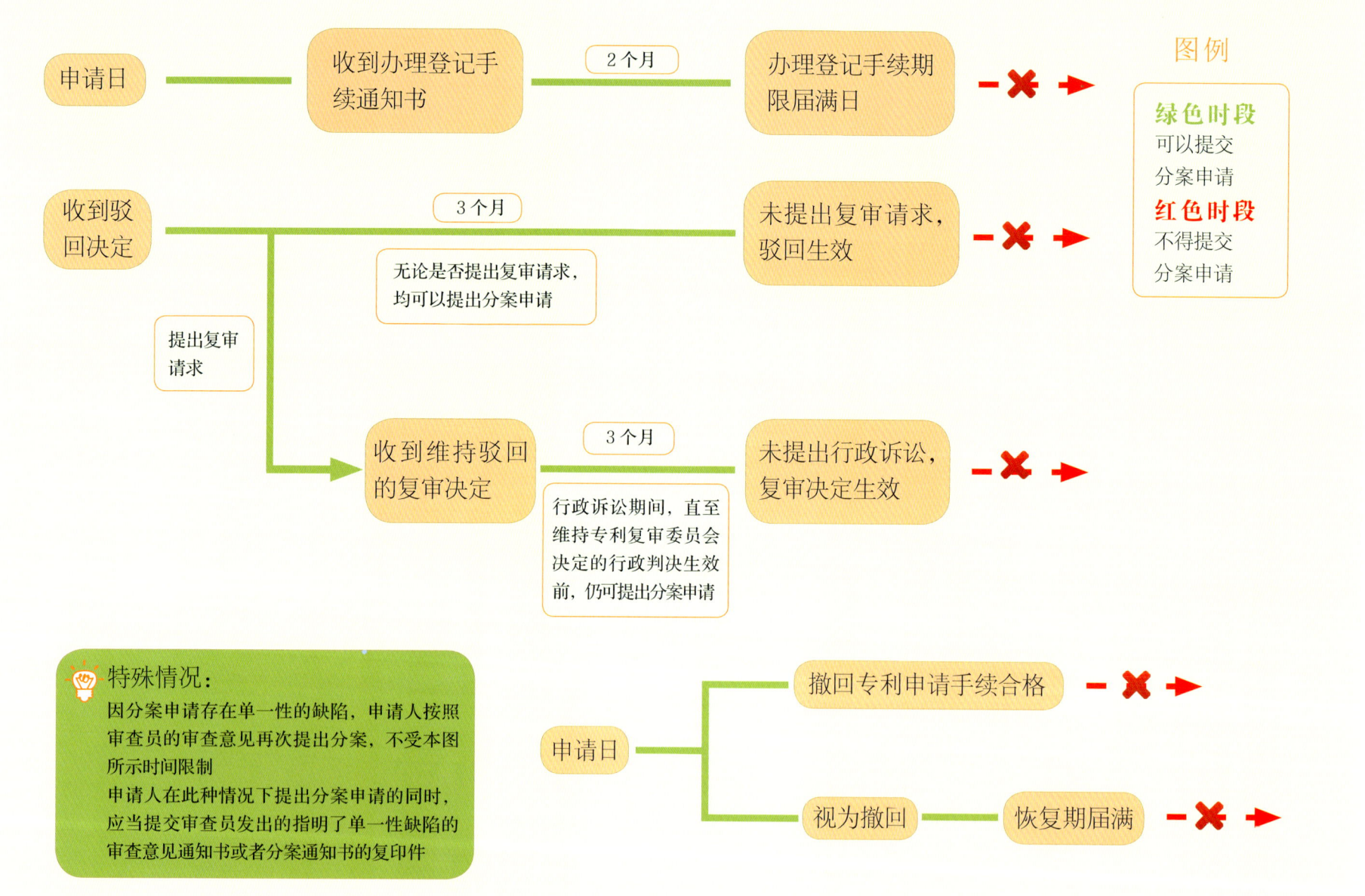

图例

绿色时段
可以提交
分案申请

红色时段
不得提交
分案申请

申请日 → 收到办理登记手续通知书 — 2个月 → 办理登记手续期限届满日

收到驳回决定 — 3个月 → 未提出复审请求，驳回生效

无论是否提出复审请求，均可以提出分案申请

提出复审请求

收到维持驳回的复审决定 — 3个月 → 未提出行政诉讼，复审决定生效

行政诉讼期间，直至维持专利复审委员会决定的行政判决生效前，仍可提出分案申请

特殊情况：
因分案申请存在单一性的缺陷，申请人按照审查员的审查意见再次提出分案，不受本图所示时间限制
申请人在此种情况下提出分案申请的同时，应当提交审查员发出的指明了单一性缺陷的审查意见通知书或者分案通知书的复印件

申请日

撤回专利申请手续合格

视为撤回 — 恢复期届满

Submission Time of Divisional Application

Filing date — Receive the notification to go through formalities of registration — [2 months] — The expiration date of going through the formalities of registration — ✖ →

Receive the decision of rejection — [3 months] — No request for reexamination, decision of rejection takes effect — ✖ →

A divisional application may be filed regardless of whether the applicant requests for reexamination or not

Request for reexamination

Receive the reexamination decision of upholding the decision of rejection — [3 months] — No request for administrative litigation, reexamination decision takes effect — ✖ →

During the initiation of the administrative litigation, until the administrative judgment of upholding the decision of Patent Reexamination Board takes effect, the applicant may also file a divisional application

Figure

Open Time

Divisional application can be filed

Close Time

No divisional application shall be filed

💡 Special Conditions :

Except that another divisional application is filed by the applicant according to the Office Action made by the examiner due to a unity defect in the divisional application

Regarding this exception, the applicant, when filing another divisional application, shall submit a copy of the Office Action indicating the unity defect or of the Notification to Make Divisional Application issued by the examiner

Filing date — Formalities of withdrawing patent application has been passed through — ✖ →

Filing date — Deemed to be withdrawn — Restoration of right expires — ✖ →

涉及生物材料专利申请的手续

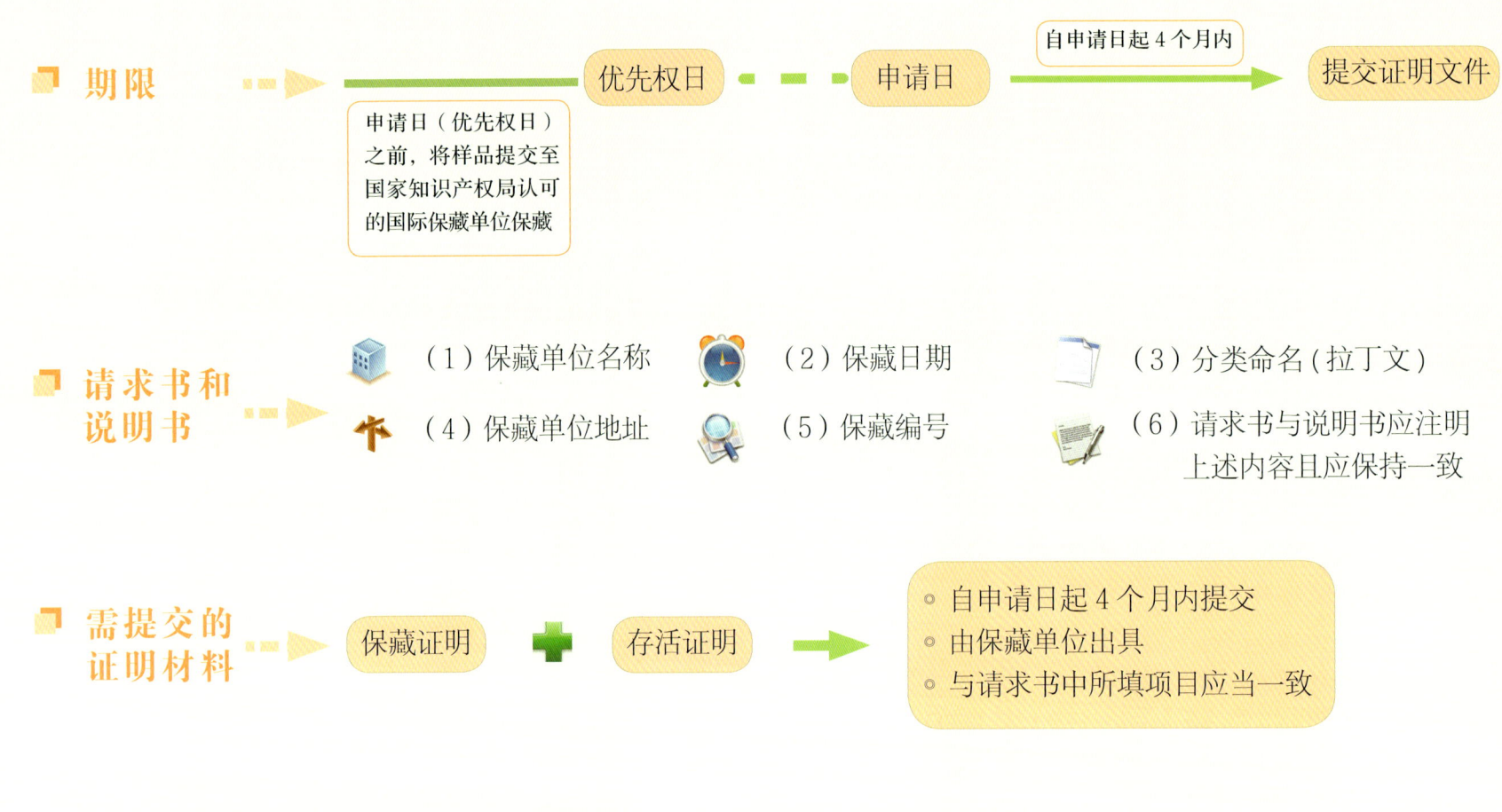

期限

优先权日 ···► 申请日 ——自申请日起4个月内——► 提交证明文件

申请日（优先权日）之前，将样品提交至国家知识产权局认可的国际保藏单位保藏

请求书和说明书

（1）保藏单位名称　（2）保藏日期　（3）分类命名（拉丁文）

（4）保藏单位地址　（5）保藏编号　（6）请求书与说明书应注明上述内容且应保持一致

需提交的证明材料

保藏证明 ＋ 存活证明 ►

- 自申请日起4个月内提交
- 由保藏单位出具
- 与请求书中所填项目应当一致

保藏日期在优先权日和申请日之间：撤回优先权要求或声明该保藏证明涉及的生物材料的内容不要求享受优先权

生物材料样品保藏手续不合格不影响初审合格，但如果申请涉及的发明必须使用的生物材料是公众不能得到的，则会导致公开不充分

Application Relating to Biological Material

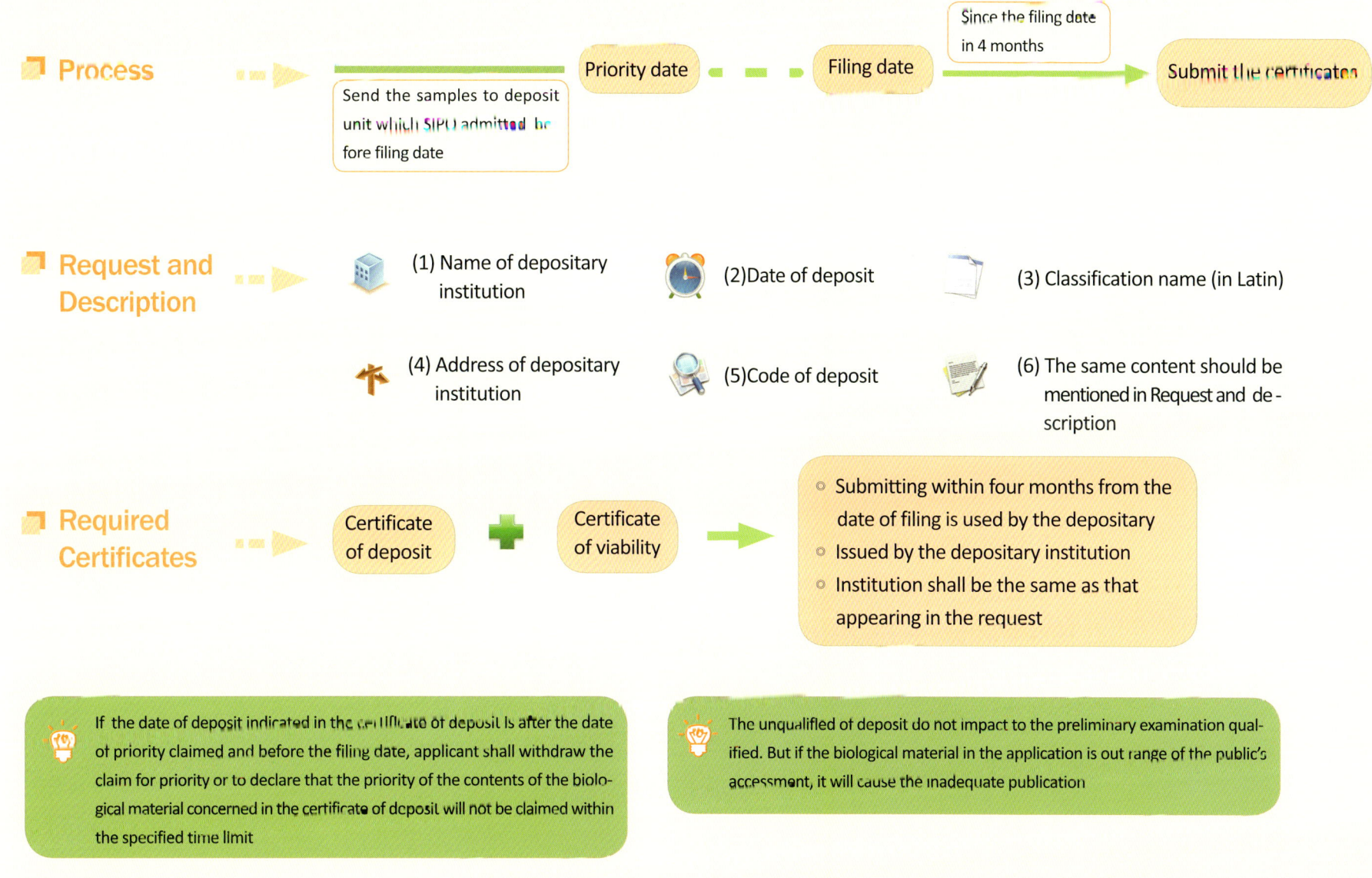

Process

Send the samples to deposit unit which SIPU admitted before filing date

Priority date

Filing date

Since the filing date in 4 months

Submit the certificaten

Request and Description

(1) Name of depositary institution

(2)Date of deposit

(3) Classification name (in Latin)

(4) Address of depositary institution

(5)Code of deposit

(6) The same content should be mentioned in Request and description

Required Certificates

Certificate of deposit

Certificate of viability

○ Submitting within four months from the date of filing is used by the depositary

○ Issued by the depositary institution

○ Institution shall be the same as that appearing in the request

If the date of deposit indicated in the certificate of deposit is after the date of priority claimed and before the filing date, applicant shall withdraw the claim for priority or to declare that the priority of the contents of the biological material concerned in the certificate of deposit will not be claimed within the specified time limit

The unqualified of deposit do not impact to the preliminary examination qualified. But if the biological material in the application is out range of the public's accessment, it will cause the inadequate publication

29

向外国申请专利保密审查手续

■ **什么情况需要提交向外申请保密审查请求** ‐‐► 任何单位或者个人将在中国完成的发明或者实用新型向外国申请专利的，应当事先报经专利局进行保密审查

■ **未经批准向外申请专利的后果**

- 在中国的专利申请不能授权
- 造成泄密的，追究责任

■ **有几种请求方式**

直接向外申请  技术方案说明书

提交国家申请时或随后向外申请 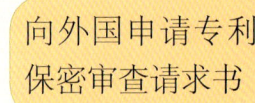 向外国申请专利保密审查请求书

 向 SIPO 提交并被受理的 PCT 申请视为同时提出向外申请保密审查请求

■ **何时可以向外国申请专利**

- 收到同意向外申请的保密审查通知或决定
- 递交日起 4 个月未收到保密审查通知或 6 个月未收到保密审查决定，默认允许

Confidentiality Examination of Patent Application to be Filed Abroad

 When to submit the request of confidentiality examination

Any entity or individual intends to file an application for patent abroad for any invention or utility model developed in China, should request the Patent Office for confidentiality examination in advance

Consequence of applying for a patent abroad without confidentiality examination

- No authorization for patent applied in China
- Shall be investigated for responsibility of the leak

Ways to request confidentiality examination

When applying for a patent abroad directly → Description of technical solution

While or after file a domestic application → Request of confidentiality examination of patent application to be filed abroad

When to file abroad

- Received the notification or decision on passing confidentiality examination
- Failed to receive the Notification on Confidentiality Examination of Patent Application to be filed abroad within 4 months or the decision on confidentiality examination of patent application to be filed abroad within 6 months

 PCT applications submitted to SIPO are deemed to request confidentiality examination

31

PCT 国际阶段流程

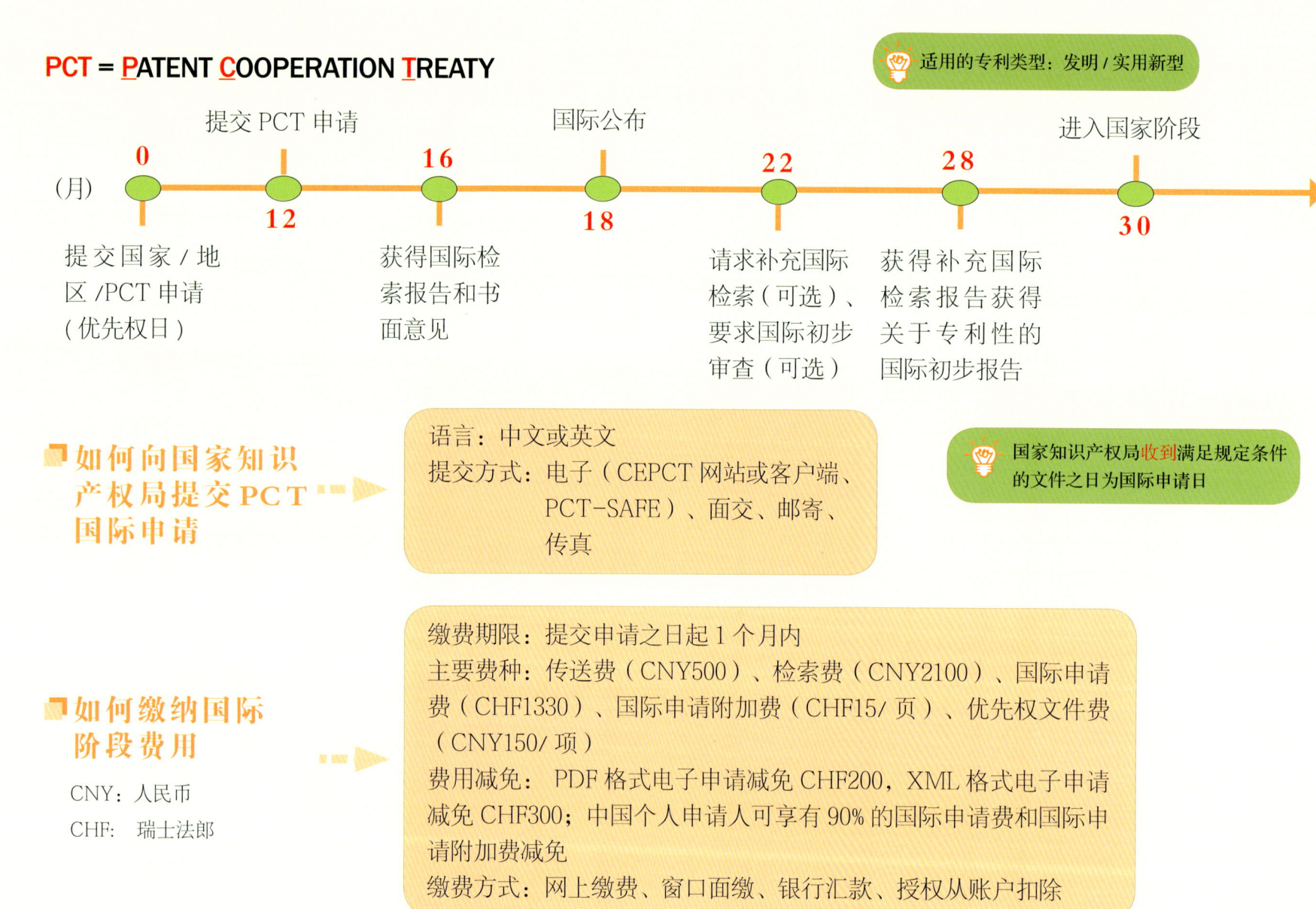

PCT = PATENT COOPERATION TREATY

适用的专利类型：发明 / 实用新型

提交 PCT 申请　　　　国际公布　　　　　　　　　　　　进入国家阶段

（月）　　0　　　　　　16　　　　　　　　22　　　　28

　　　　　12　　　　　　　　18　　　　　　　　　　　　　30

提交国家 / 地区 /PCT 申请（优先权日）

获得国际检索报告和书面意见

请求补充国际检索（可选）、要求国际初步审查（可选）

获得补充国际检索报告获得关于专利性的国际初步报告

如何向国家知识产权局提交 PCT 国际申请

语言：中文或英文
提交方式：电子（CEPCT 网站或客户端、PCT-SAFE）、面交、邮寄、传真

国家知识产权局收到满足规定条件的文件之日为国际申请日

如何缴纳国际阶段费用

CNY：人民币
CHF： 瑞士法郎

缴费期限：提交申请之日起 1 个月内
主要费种：传送费（CNY500）、检索费（CNY2100）、国际申请费（CHF1330）、国际申请附加费（CHF15/ 页）、优先权文件费（CNY150/ 项）
费用减免： PDF 格式电子申请减免 CHF200，XML 格式电子申请减免 CHF300；中国个人申请人可享有 90% 的国际申请费和国际申请附加费减免
缴费方式：网上缴费、窗口面缴、银行汇款、授权从账户扣除

PCT International Phase

PCT = **P**ATENT **C**OOPERATION **T**REATY

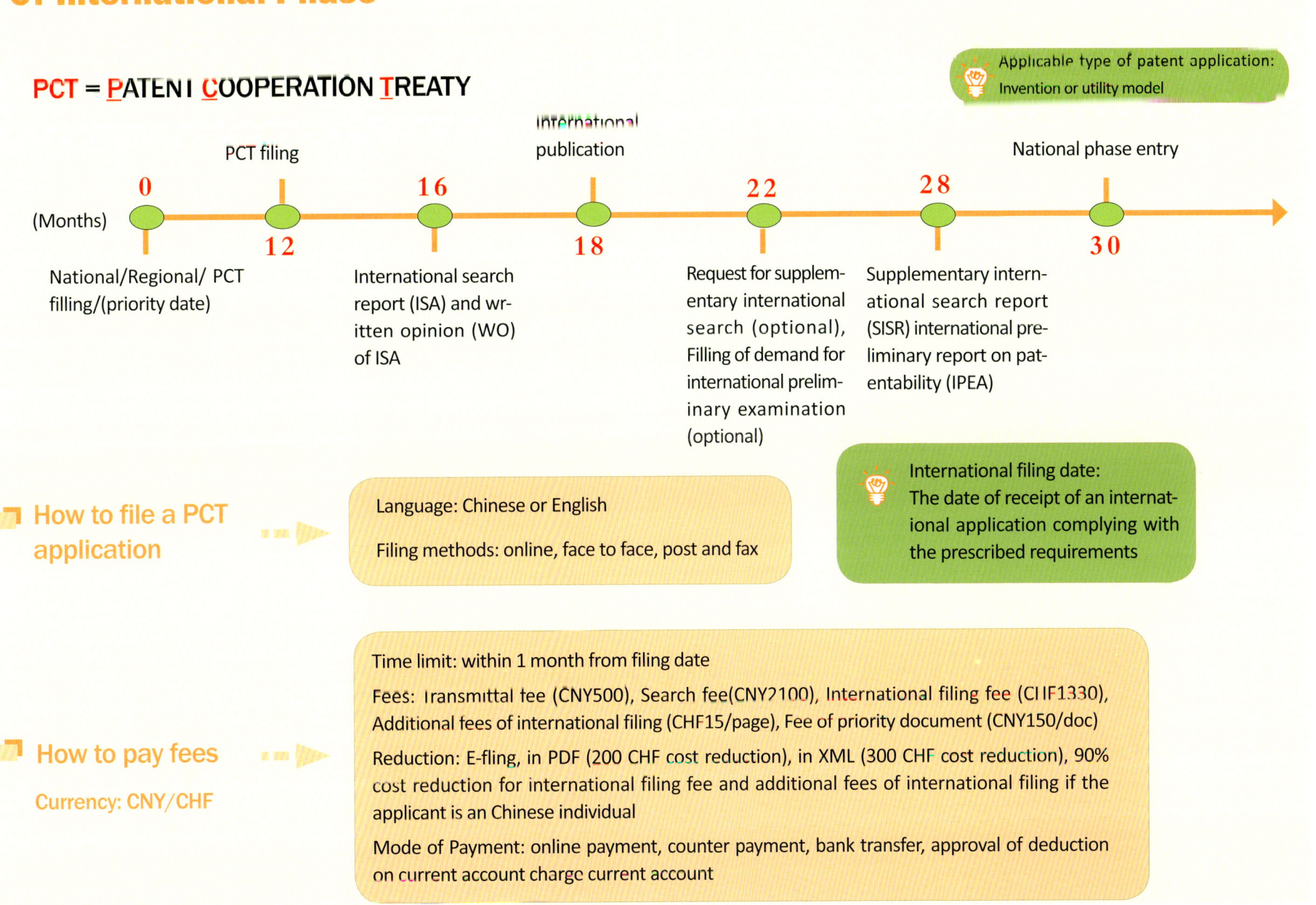

Applicable type of patent application: Invention or utility model

PCT filing

International publication

National phase entry

0   16   22   28

(Months)   12   18   30

National/Regional/ PCT filling/(priority date)

International search report (ISA) and written opinion (WO) of ISA

Request for supplementary international search (optional), Filling of demand for international preliminary examination (optional)

Supplementary international search report (SISR) international preliminary report on patentability (IPEA)

International filing date: The date of receipt of an international application complying with the prescribed requirements

How to file a PCT application

Language: Chinese or English

Filing methods: online, face to face, post and fax

How to pay fees

Currency: CNY/CHF

Time limit: within 1 month from filing date

Fees: Transmittal fee (CNY500), Search fee(CNY2100), International filing fee (CHF1330), Additional fees of international filing (CHF15/page), Fee of priority document (CNY150/doc)

Reduction: E-fling, in PDF (200 CHF cost reduction), in XML (300 CHF cost reduction), 90% cost reduction for international filing fee and additional fees of international filing if the applicant is an Chinese individual

Mode of Payment: online payment, counter payment, bank transfer, approval of deduction on current account charge current account

要求 PCT 国际申请优先权

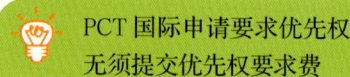

PCT 国际
申请如何要
求优先权

在提交申请时或规定的期限内提出优先权要求

在先申请的申请日

在先申请的申请号

在先申请的受理机构

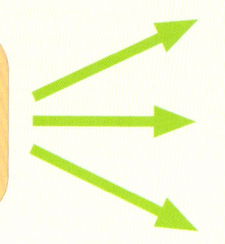

如何提交优先权文件

提交方式	提交期限
直接提交经在先申请的受理机构证明的优先权文件	向受理局提交：优先权日起 16 个月内 向国际局提交：国际公布前
如果优先权文件由受理局出具，可以请求受理局准备优先权文件并将该文件送交国际局，但需缴纳优先权文件费	优先权日起 16 个月内向受理局提交请求
提供优先权文件的 DAS 查询码，请国际局从数字图书馆获取优先权文件	在国际公布前请求国际局通过 DAS 获取优先权文件；保证优先权文件在国际公布前能通过 DAS 被国际局获得

Priority Claims in PCT International Phase

■ **Claiming priority in the PCT international phase**

→ Claiming priority at the time of filing a PCT application or within the prescribed time limit

→ The filing date of the Earlier Application

→ Application number of the prior application

→ The receiving office of the prior application

 It is free of charge to claim priority in a PCT application

■ **How to submit priority documents**

HOW	WHEN
Furnish such priority document directly to SIPO or the International Bureau (IB)	Direct submission to SIPO: within 16 months from the priority date. Direct submission to IB: before international publication
Request SIPO to prepare such priority document and transmit to the IB , if the prior application was filed with SIPO	Within 16 months from the priority date
Request IB to obtain such priority document from a digital library	Such priority document must be made available to the IB via DAS and the request to the IB to retrieve the priority document must be made before international publication

办理 PCT 申请进入中国国家阶段的准备

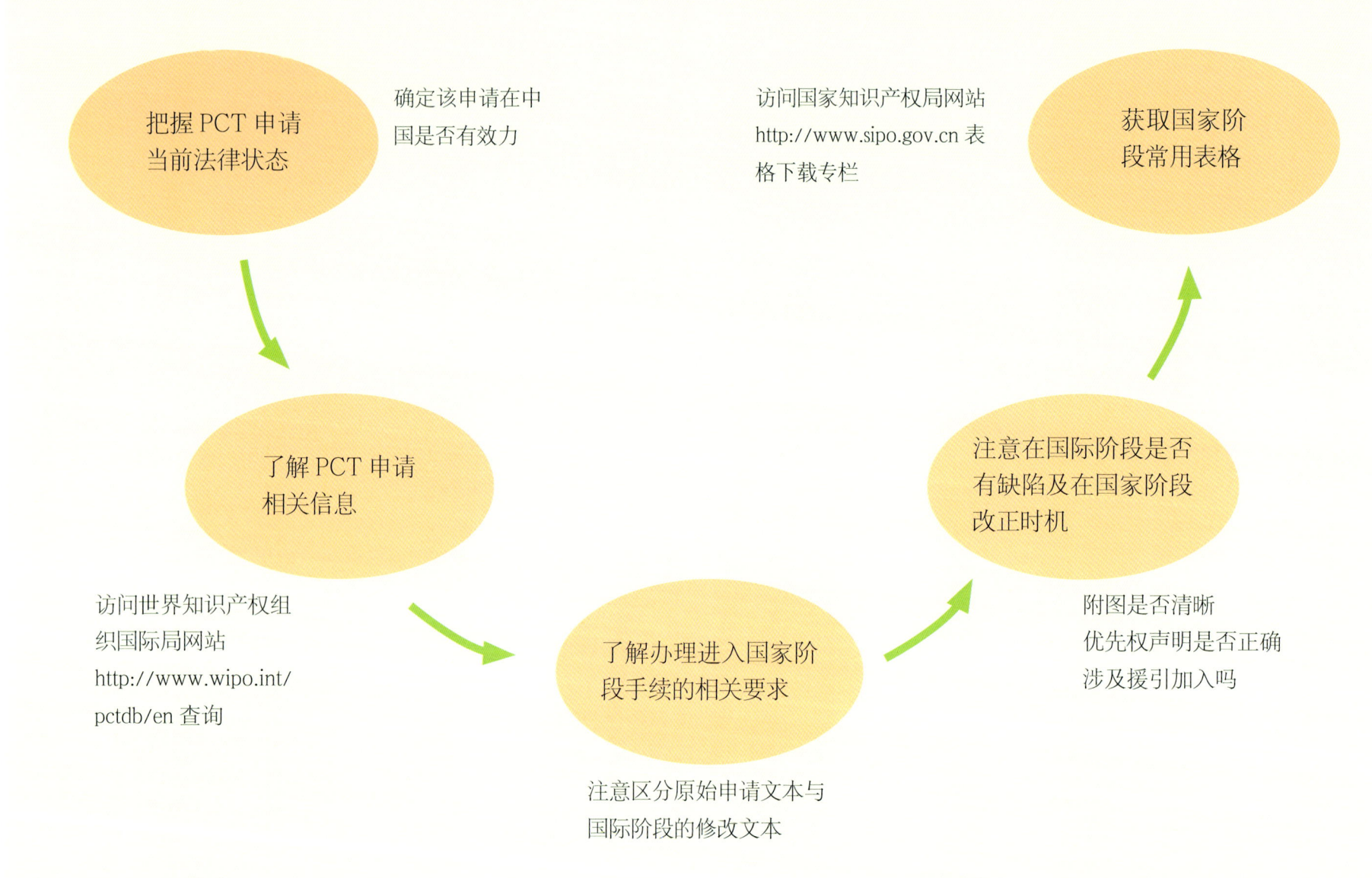

把握 PCT 申请
当前法律状态

确定该申请在中
国是否有效力

访问国家知识产权局网站
http://www.sipo.gov.cn 表
格下载专栏

获取国家阶
段常用表格

了解 PCT 申请
相关信息

访问世界知识产权组
织国际局网站
http://www.wipo.int/
pctdb/en 查询

了解办理进入国家阶
段手续的相关要求

注意区分原始申请文本与
国际阶段的修改文本

注意在国际阶段是否
有缺陷及在国家阶段
改正时机

附图是否清晰
优先权声明是否正确
涉及援引加入吗

The Preparation for the PCT Application National Phase

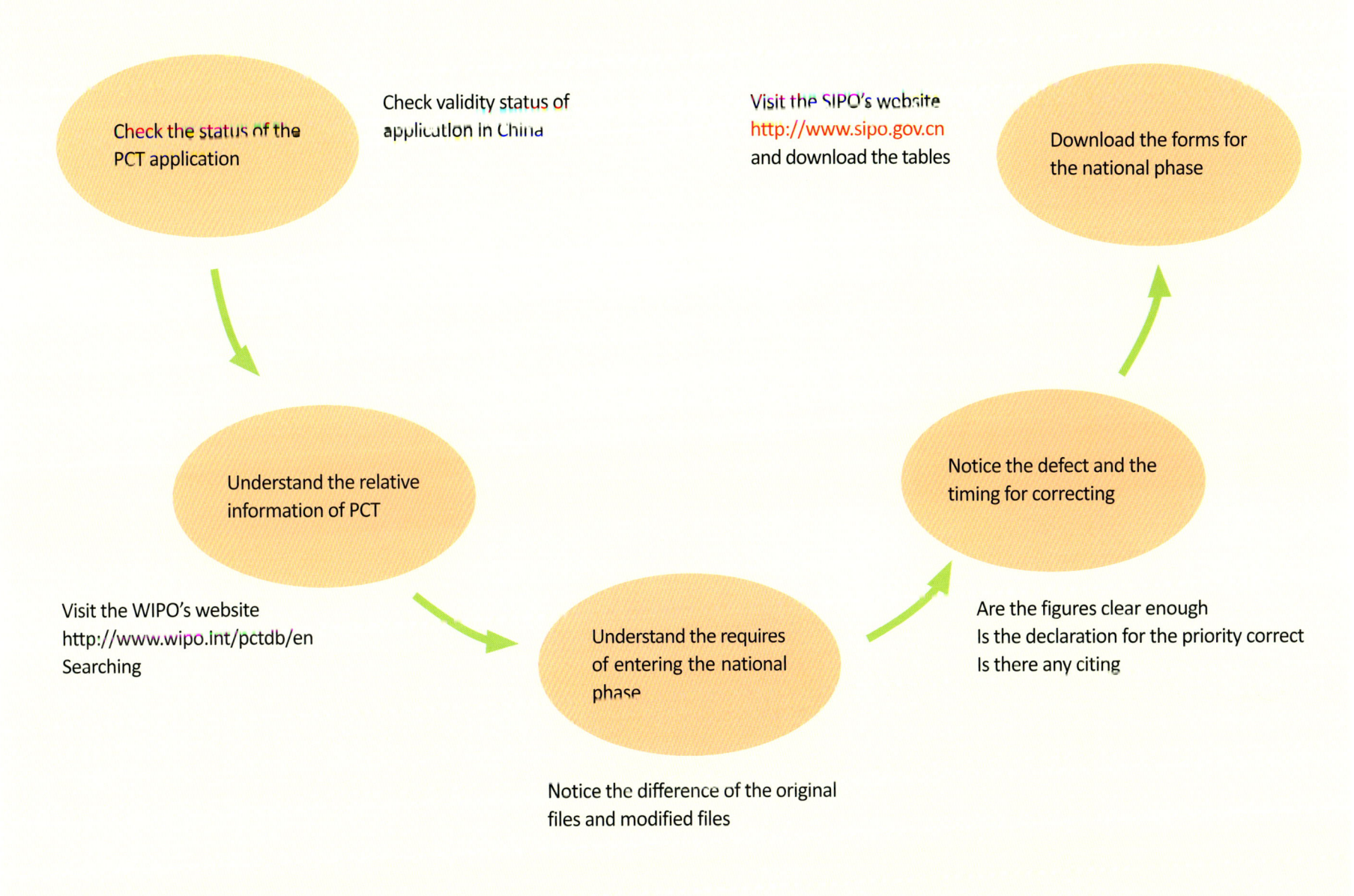

Check the status of the PCT application

Check validity status of application in China

Visit the SIPO's website
http://www.sipo.gov.cn
and download the tables

Download the forms for the national phase

Understand the relative information of PCT

Visit the WIPO's website
http://www.wipo.int/pctdb/en
Searching

Understand the requires of entering the national phase

Notice the difference of the original files and modified files

Notice the defect and the timing for correcting

Are the figures clear enough
Is the declaration for the priority correct
Is there any citing

PCT 国际申请在中国的效力

在中国有效力的情形

经受理局审查符合《专利合作条约》第11条的规定、确定了国际申请日，并且指定了中国的国际申请，该申请具有正规的国家申请效力

在中国没有效力的情形

声称进入国家阶段，国际公布文本中没有指定中国的记载的国际申请（因对中国的指定被撤回）

在中国的效力丧失的情形

（1）国际阶段：国际申请被撤回、国际申请被视为撤回
（2）未在期限内办理进入国家阶段手续
（3）自优先权之日起32个月期限届满时进入国家阶段手续仍不满足最低要求

 效力的重要性

国际申请在中国没有效力或者在中国的效力丧失的，不能进入国家阶段

The Effect of PCT International Application in China

⅂ Having effect in China

The application which has passed the examination by the receiving Office on whether it conforms with Article 11 of the Treaty, been accorded a filing date and designated China has effect as a regular national application

⅂ Having no effect In China

Analleged international application entering the national phase without indication of the designation of China in its international publication document (or the designation has been withdrawn)

⅂ Losing effect in China

(1) International Phase: the international application has been withdrawn or deemed withdrawn
(2) Fails to go through the formalities for entering the national phase within the time limit
(3) Does not meet the minimum requirements for entering the national phase even after 32 months from the priority date

 The Importance of Effect

International application with effect or loss of the effect in China cannot enter the national phase

39

PCT 国际申请进入国家阶段手续

提交什么文件

外文提出的 PCT 国际申请
- 进入声明：国际申请号、保护类型
- 说明书中译文、权利要求书中译文、摘要中译文、说明书附图中文副本、 摘要附图中文副本等

中文提出的 PCT 国际申请
- 进入声明：国际申请号、保护类型
- 摘要、摘要附图副本等

何时办理

优先权日起 30 个月内，缴纳宽限费延长至 32 个月

怎么缴纳费用

非中国局受理的 PCT 国际申请
发明：申请费、公布印刷费、申请附加费、宽限费
实用新型：申请费、申请附加费、宽限费
中国局受理的 PCT 国际申请
发明：公布印刷费、宽限费
实用新型：宽限费

进入日的确定

办理的进入国家阶段的手续满足最低要求之日，以文件递交日和费用缴纳日两者后到日为准

红色字体内容为进入手续最低要求；
加下画线的内容为适用时提交 / 缴纳

The Formalities of PCT Application in National Phase

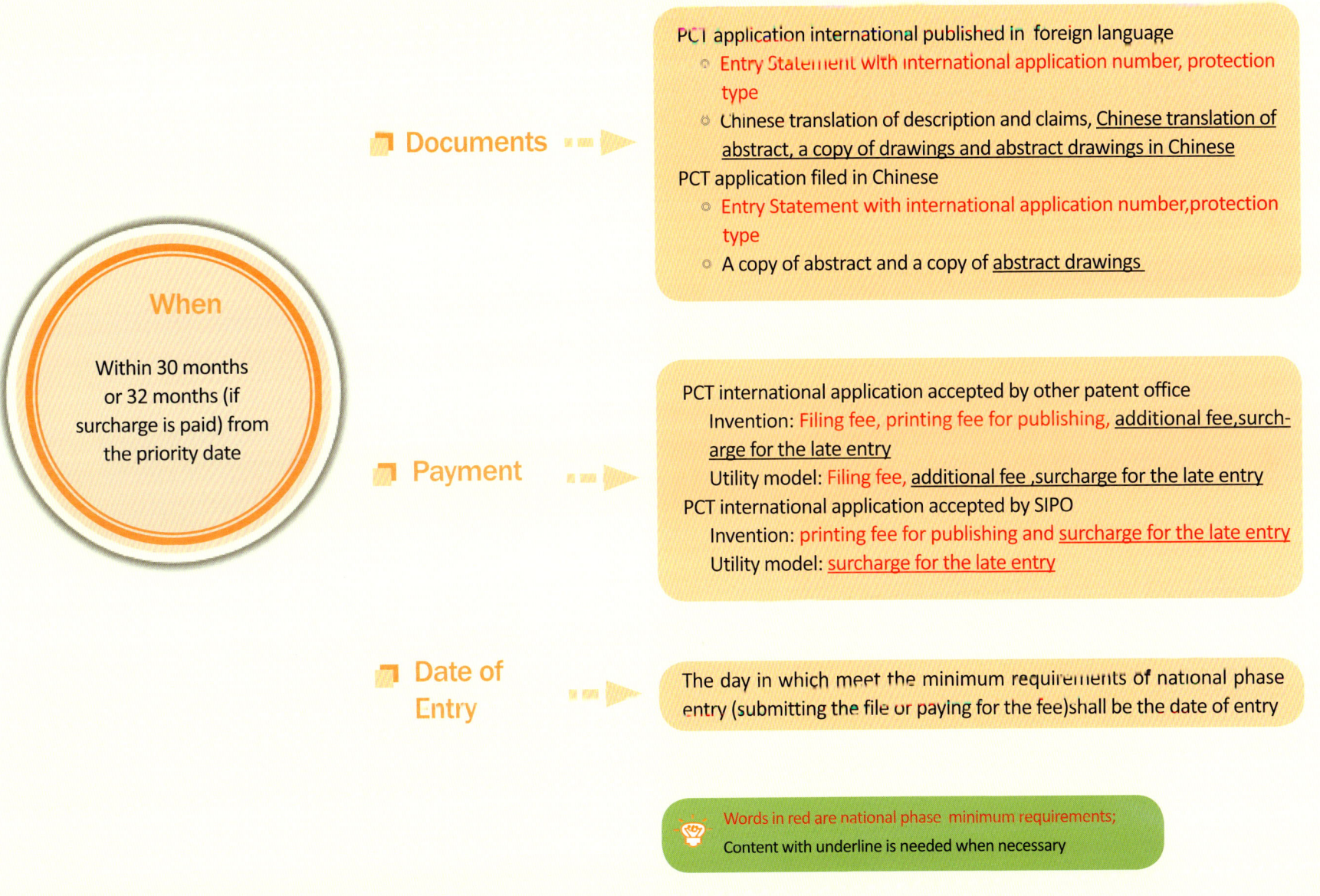

When

Within 30 months
or 32 months (if
surcharge is paid) from
the priority date

Documents

PCT application international published in foreign language
- Entry Statement with international application number, protection type
- Chinese translation of description and claims, Chinese translation of abstract, a copy of drawings and abstract drawings in Chinese

PCT application filed in Chinese
- Entry Statement with international application number, protection type
- A copy of abstract and a copy of abstract drawings

Payment

PCT international application accepted by other patent office
    Invention: Filing fee, printing fee for publishing, additional fee, surcharge for the late entry
    Utility model: Filing fee, additional fee, surcharge for the late entry
PCT international application accepted by SIPO
    Invention: printing fee for publishing and surcharge for the late entry
    Utility model: surcharge for the late entry

Date of Entry

The day in which meet the minimum requirements of national phase entry (submitting the file or paying for the fee)shall be the date of entry

Words in red are national phase minimum requirements;
Content with underline is needed when necessary

PCT 国际申请在中国国家阶段改正译文错误

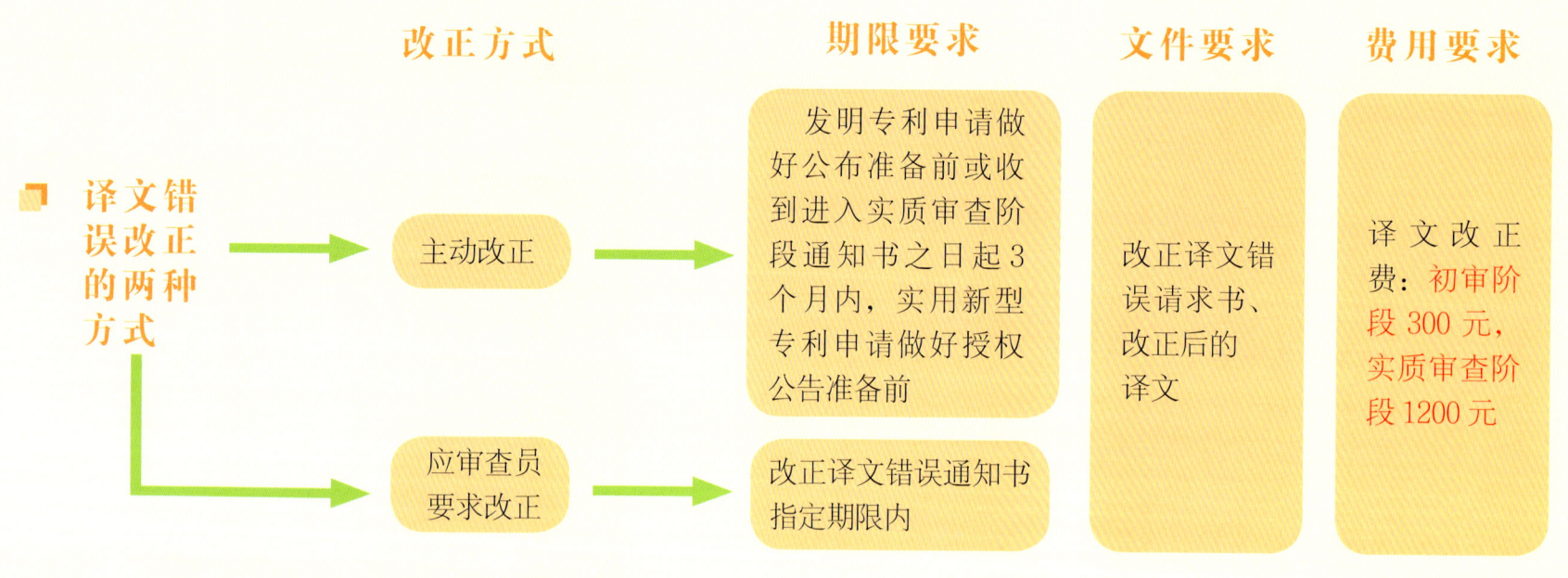

■ 什么是译文错误 ⟶ 说明书、权利要求书或说明书附图中文字的译文与原文相比,个别术语、个别句子或者个别段落遗漏或不准确

■ 译文错误改正的两种方式

改正方式	期限要求	文件要求	费用要求
主动改正	发明专利申请做好公布准备前或收到进入实质审查阶段通知书之日起3个月内,实用新型专利申请做好授权公告准备前	改正译文错误请求书、改正后的译文	译文改正费:初审阶段300元,实质审查阶段1200元
应审查员要求改正	改正译文错误通知书指定期限内		

中文译文与原文明显不相符的,不属于译文错误

Correction of Translation Errors for PCT International Application Entering National Phase

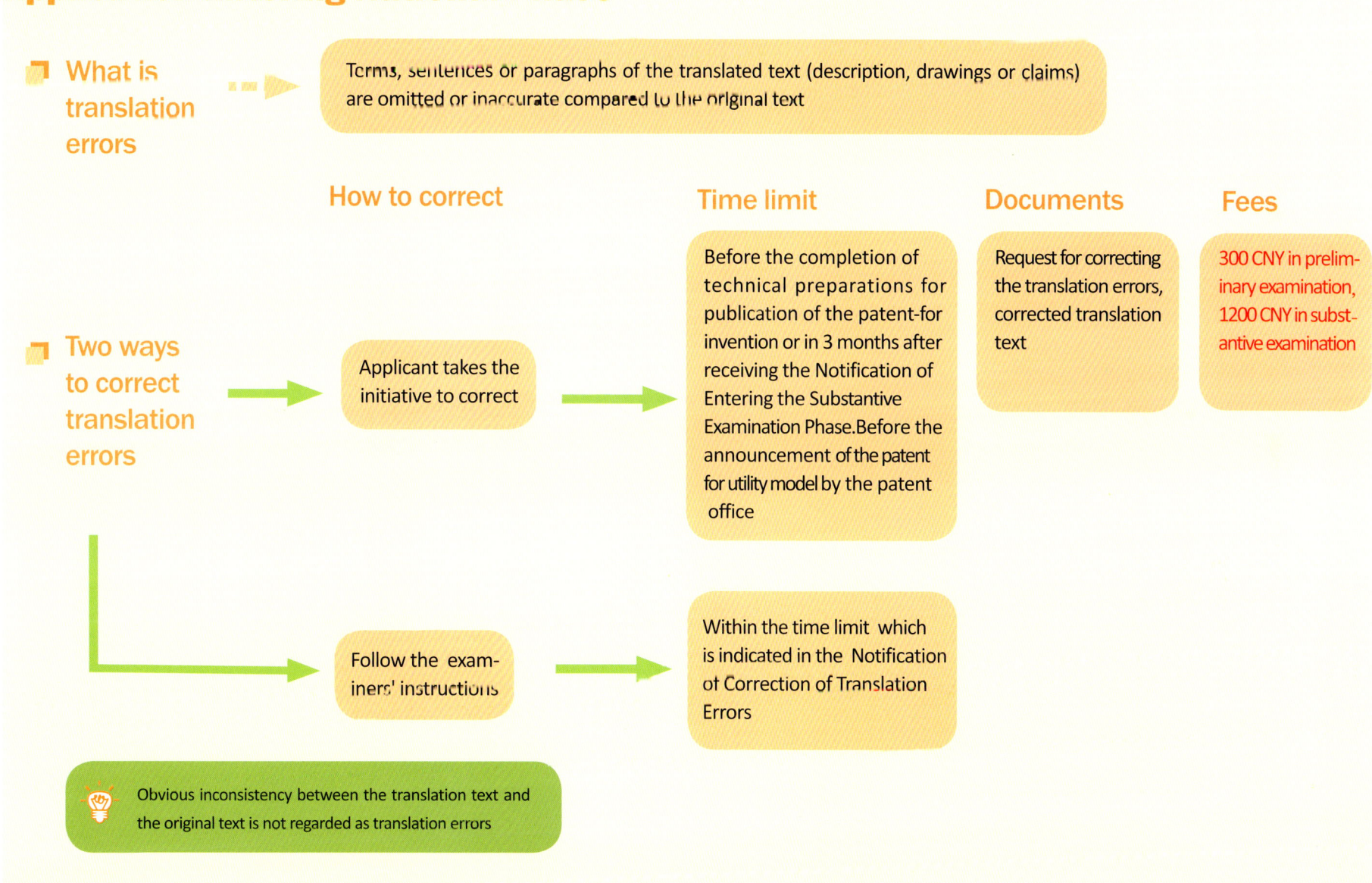

What is translation errors

Terms, sentences or paragraphs of the translated text (description, drawings or claims) are omitted or inaccurate compared to the original text

Two ways to correct translation errors

How to correct	Time limit	Documents	Fees
Applicant takes the initiative to correct	Before the completion of technical preparations for publication of the patent-for invention or in 3 months after receiving the Notification of Entering the Substantive Examination Phase.Before the announcement of the patent for utility model by the patent office	Request for correcting the translation errors, corrected translation text	300 CNY in preliminary examination, 1200 CNY in substantive examination
Follow the examiners' instructions	Within the time limit which is indicated in the Notification of Correction of Translation Errors		

Obvious inconsistency between the translation text and the original text is not regarded as translation errors

发明专利优先审查手续

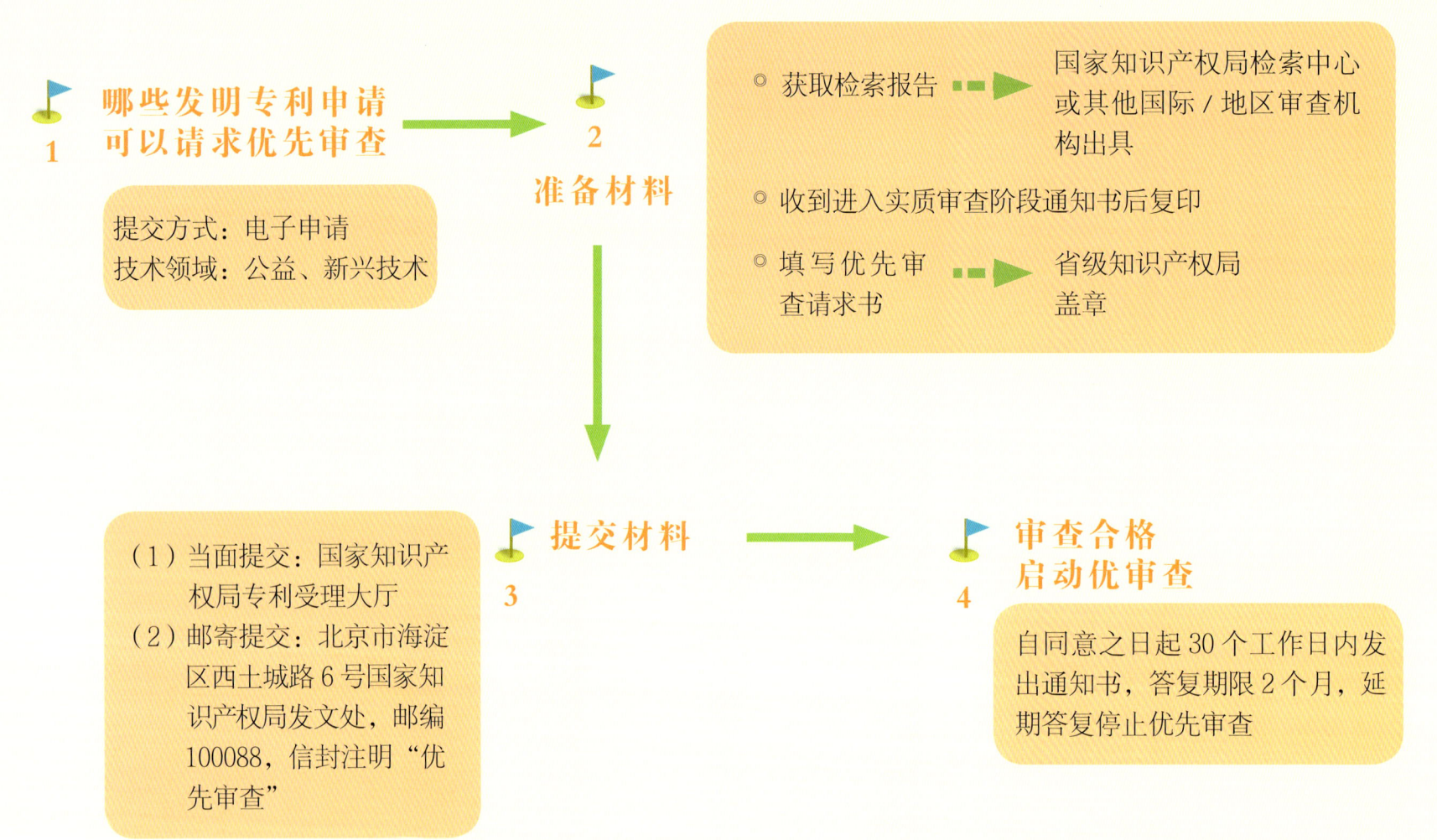

1 哪些发明专利申请可以请求优先审查

提交方式：电子申请
技术领域：公益、新兴技术

2 准备材料

◎ 获取检索报告 ➡ 国家知识产权局检索中心或其他国际／地区审查机构出具

◎ 收到进入实质审查阶段通知书后复印

◎ 填写优先审查请求书 ➡ 省级知识产权局盖章

3 提交材料

（1）当面提交：国家知识产权局专利受理大厅
（2）邮寄提交：北京市海淀区西土城路6号国家知识产权局发文处，邮编100088，信封注明"优先审查"

4 审查合格 启动优审查

自同意之日起30个工作日内发出通知书，答复期限2个月，延期答复停止优先审查

The Procedure of Patent Earlier Examination

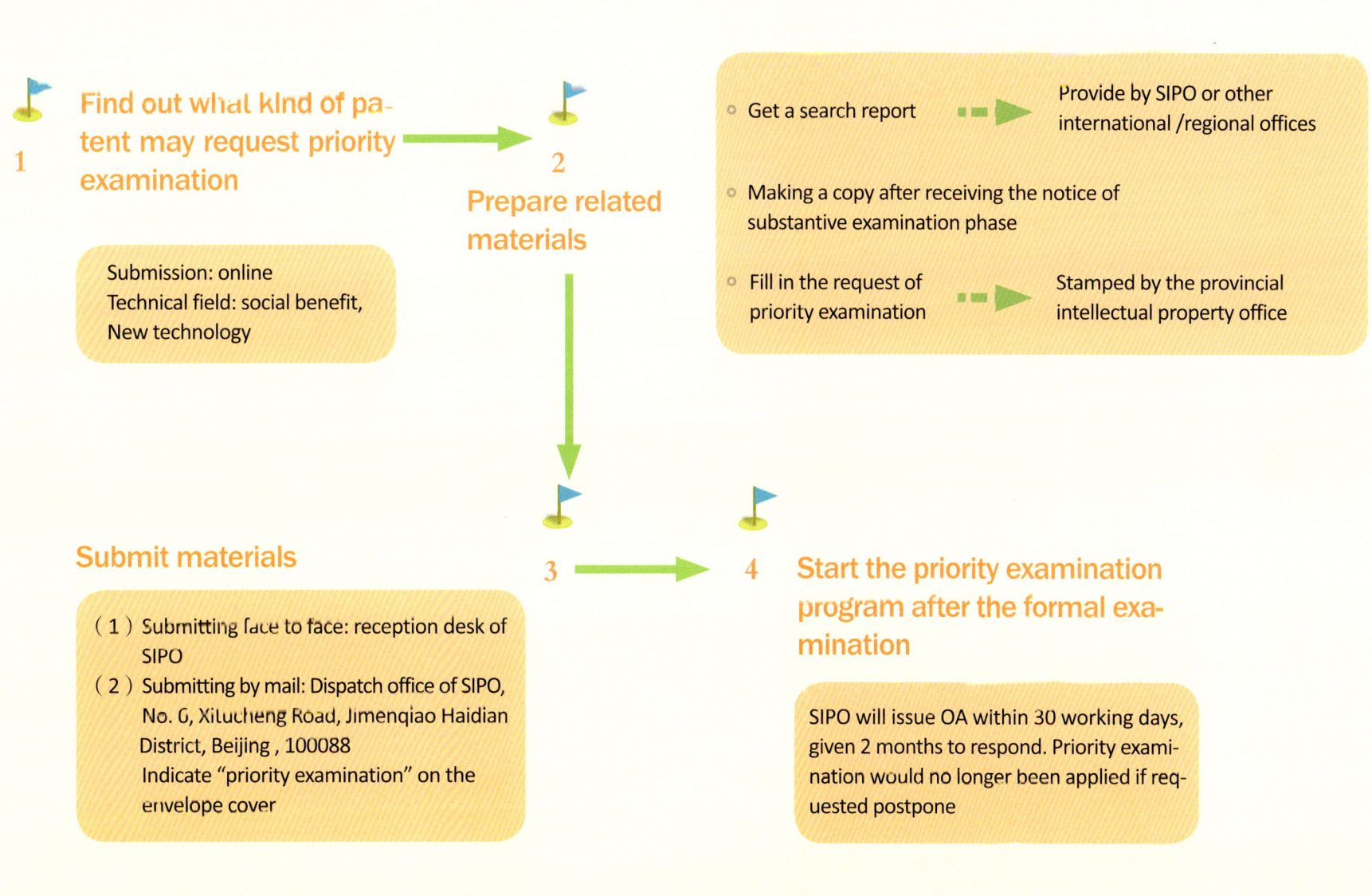

1 Find out what kind of patent may request priority examination

Submission: online
Technical field: social benefit,
New technology

2 Prepare related materials

○ Get a search report ➡ Provide by SIPO or other international /regional offices

○ Making a copy after receiving the notice of substantive examination phase

○ Fill in the request of priority examination ➡ Stamped by the provincial intellectual property office

3 Submit materials

（1）Submitting face to face: reception desk of SIPO

（2）Submitting by mail: Dispatch office of SIPO, No. 6, Xitucheng Road, Jimenqiao Haidian District, Beijing , 100088
Indicate "priority examination" on the envelope cover

4 Start the priority examination program after the formal examination

SIPO will issue OA within 30 working days, given 2 months to respond. Priority examination would no longer been applied if requested postpone

45

专利审查高速路（PPH）

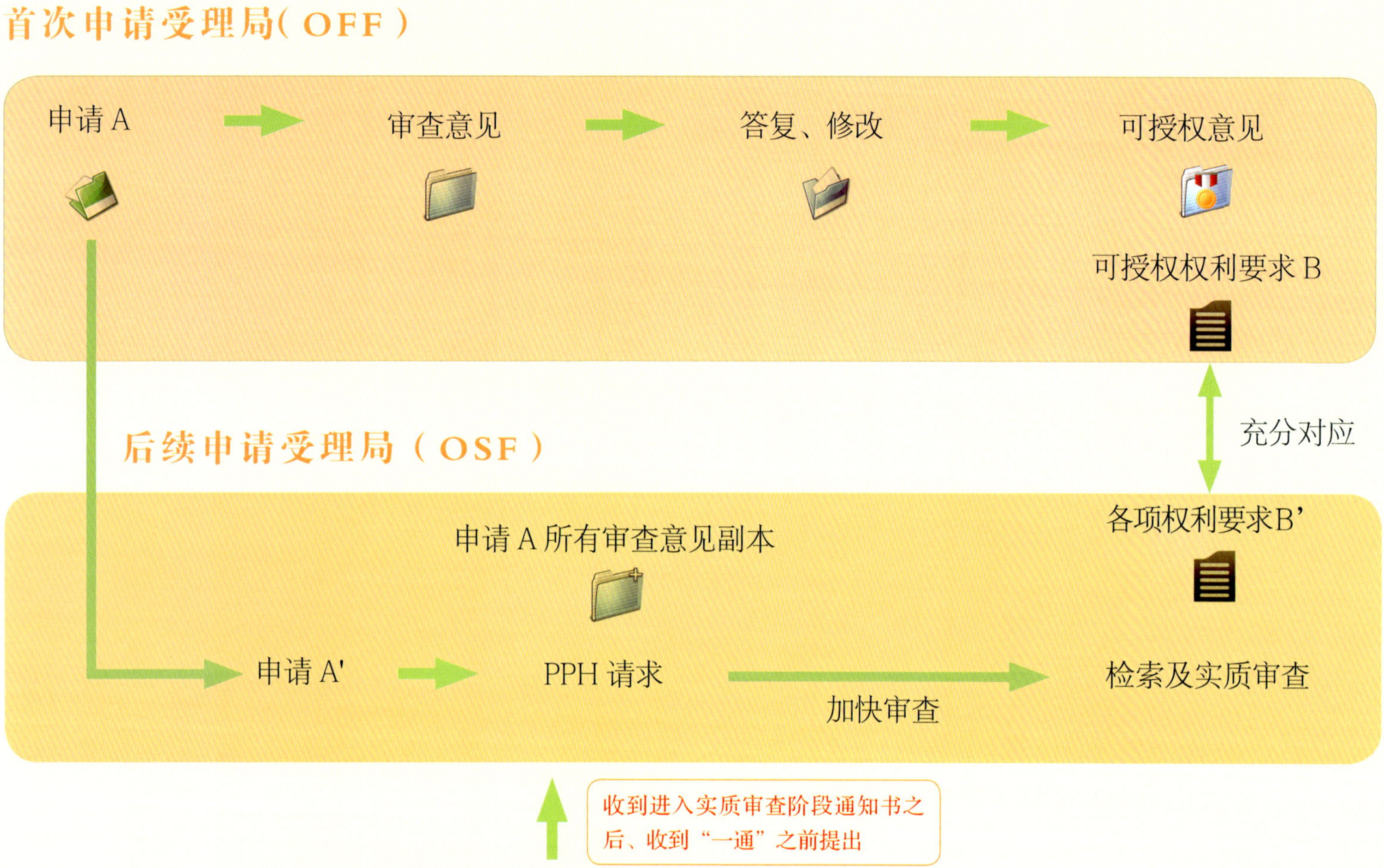

首次申请受理局（OFF）

申请 A ➡ 审查意见 ➡ 答复、修改 ➡ 可授权意见

可授权权利要求 B

充分对应

后续申请受理局（OSF）

申请 A 所有审查意见副本

各项权利要求 B'

申请 A' ➡ PPH 请求 —— 加快审查 ——➡ 检索及实质审查

收到进入实质审查阶段通知书之后、收到"一通"之前提出

两种类型

常规 PPH：利用《巴黎公约》路径或 PCT 国家阶段的 OFF 作出的国内工作结果提出
PCT-PPH：利用 PCT 国际阶段工作结果提出

 扩展：五局多边合作 PPH

Patent Prosecution Highway(PPH)

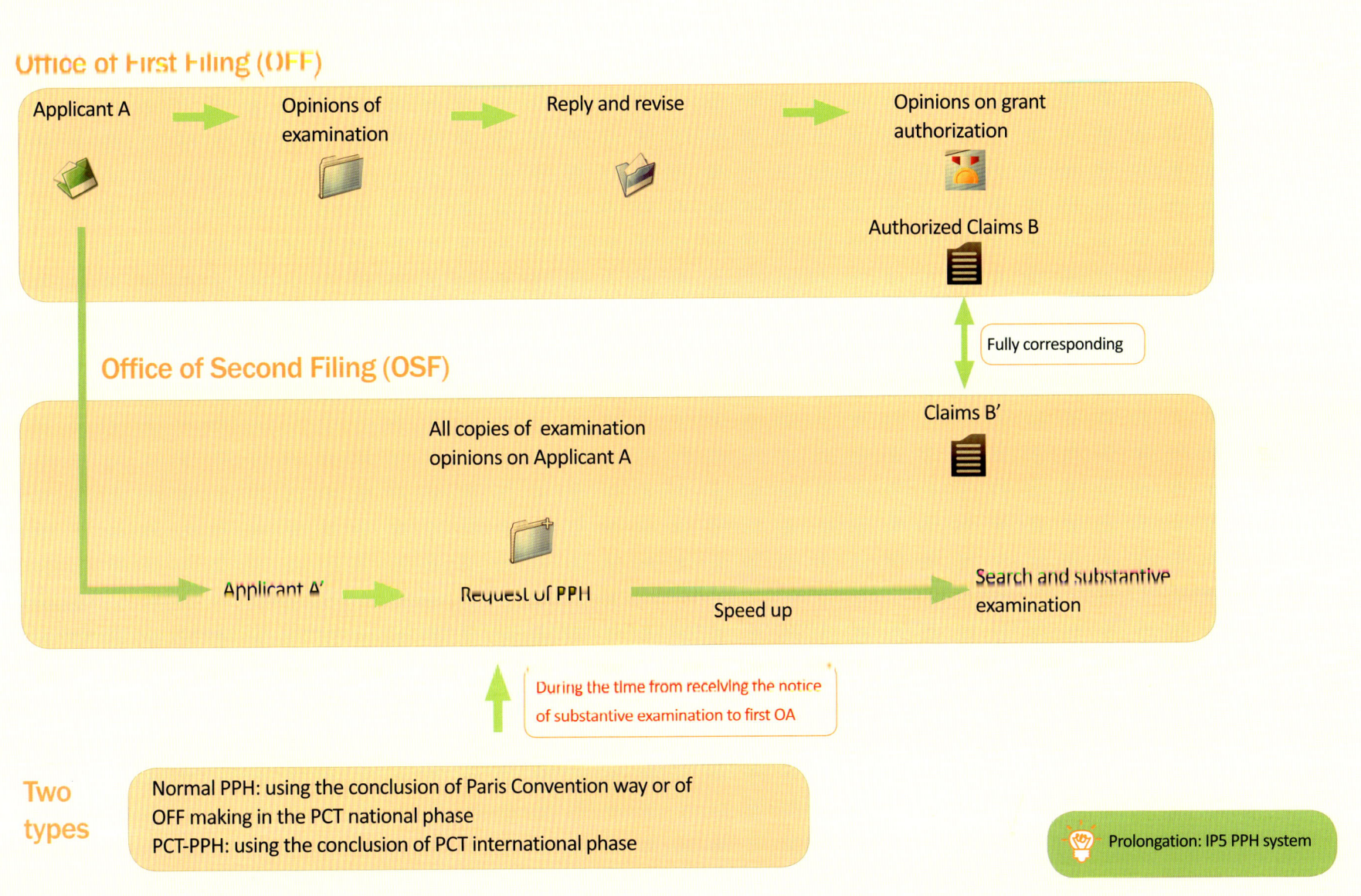

Office of First Filing (OFF)

Applicant A → Opinions of examination → Reply and revise → Opinions on grant authorization

Authorized Claims B

Fully corresponding

Office of Second Filing (OSF)

All copies of examination opinions on Applicant A

Claims B'

Applicant A' → Request of PPH → Speed up → Search and substantive examination

During the time from receiving the notice of substantive examination to first OA

Two types

Normal PPH: using the conclusion of Paris Convention way or of OFF making in the PCT national phase
PCT-PPH: using the conclusion of PCT international phase

Prolongation: IP5 PPH system

主动修改和依职权修改

主动修改的时机

 发明专利申请
 发明专利国际申请
→ 提出实质审查请求时或收到发明专利申请进入实质审查阶段通知书之日起的 3 个月内

 实用新型专利申请 → 自申请日起 2 个月内
 实用新型专利国际申请 → 自进入日起 2 个月内

 外观设计专利申请 → 自申请日起 2 个月内

主动修改的范围

◎ 发明或实用新型专利申请
对申请文件的修改不得超出原说明书和权利要求书记载的范围
◎ 外观设计专利申请
对申请文件的修改不得超出原图片或照片记载的范围

依职权修改的范围 → 修改专利申请文件中文字和符号的明显错误

依职权修改的告知 → 专利局对申请文件的依职权修改会通知申请人

The Amendment by the Applicant or the Examiner

🗂 Time to submit the amendment

 Applications of invention

 Applications of invention through PCT

(1) at the time when a request for examination as to substance is made

(2) when within the time limit of 3 months after the receipt of the notification of the Patent Office on the entry into examination as to substance of the application

 Applications of utility model

 2 months since the filing date

 Applications of utility model through PCT

2 months since the entering date

 Applications for design

 2 months since the filing date

🗂 The range of the amendment by examiner

make amendment to the obvious clerical mistakes and symbol mistakes

🗂 The notification of the amendment by examiner

 the applicant will be notified for the amendment ex officio made by examiner

🗂 Range of the initiative amendment

○ **Invention or utility model**
the amendments should not go beyond the scope described in the initial description and claims

○ **Design**
the amendments should not go beyond the scope of the disclosure as shown in the initial drawings or photographs

49

著录项目变更手续

■ **著录项目变更手续主要针对哪些信息** ⇒ 申请人或专利权人 ➕ 代表人 ➕ 联系人 ➕ 发 明 人（更正） ➕ 代理机构及代理人

■ **谁办理著录项目变更手续** ⇒ 未委托专利代理机构的，由申请人（或专利权人）或其代表人办理；已委托专利代理机构的，由专利代理机构办理。因权利转移引起的变更，也可以由新的权利人或者其委托的专利代理机构办理

■ **如何办理著录项目变更手续** ⇒ 著录项目变更申报书 **正确填写变更前、变更后信息，签章**

著录项目变更证明文件
◎ **转让：转让证明**
◎ **企业更名：工商部门证明**
◎ **发明人更正：变更前全体申请人和发明人同意的证明**

变更费（1个月内缴纳）
◎ **申请人或专利权人，发明人：200元**
◎ **代理机构或代理人：50元**
◎ **一次变更最多250元**

💡 电子申请的受让方需提前注册电子申请用户

💡 手续合格通知书发文日起变更生效

Bibliographic Data Change

What is bibli-ographic data ┄▶ Applicant or patent owner ➕ Representative ➕ Contact person ➕ Inventor (correction) ➕ Agent and Agency

Who is auth-orized to change bibl-iographic data ┄▶ If no patent agency is appointed, the applicant (or patentee), or his representative should go through the formalities; otherwise, the formalities shall be gone through by the patent agency. The forma-lities may be gone through by the new right owner or his appointed patent agency if the change is due to the transfer of the right

How to change bibliographic data ┄▶

Statement for change in bibliographic data
Filled in correct information of "before change" and "after change" with sign-ature

Certifying document for change in bibliographic data
○ Transfer of patent right : agreement
○ Renaming of company:certificate given by industry and commerce department
○ Change of inventor: certificate with sign atures of all applicants and inventors be-fore the change

Alteration fee(should be paid within 1 month)
○ Applicant, patent owner or inventor: 200 CNY
○ Agent or agency: 50 CNY
○ Max 250 CNY in one request of change in bibliographic data

 Assignee of an electronic application should register an electronic account online first

 The change in the bibliographic data shall take affect from the date of issuance of Notification of Passing Examination of Formalities by the patent office

51

发明专利公布程序

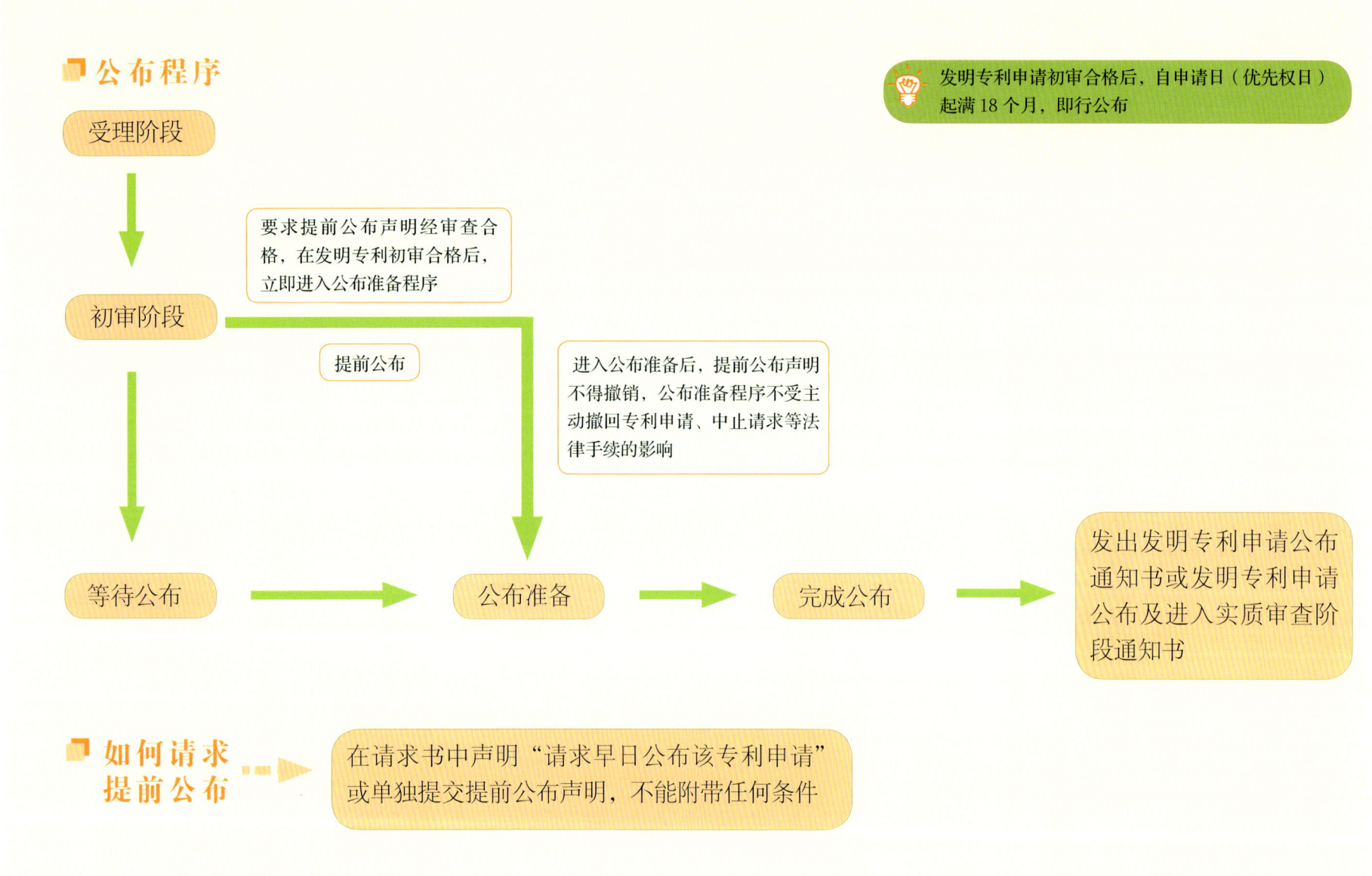

公布程序

受理阶段

初审阶段

要求提前公布声明经审查合格，在发明专利初审合格后，立即进入公布准备程序

提前公布

进入公布准备后，提前公布声明不得撤销，公布准备程序不受主动撤回专利申请、中止请求等法律手续的影响

发明专利申请初审合格后，自申请日（优先权日）起满18个月，即行公布

等待公布

公布准备

完成公布

发出发明专利申请公布通知书或发明专利申请公布及进入实质审查阶段通知书

如何请求提前公布

在请求书中声明"请求早日公布该专利申请"或单独提交提前公布声明，不能附带任何条件

Publication of the Application for Invention

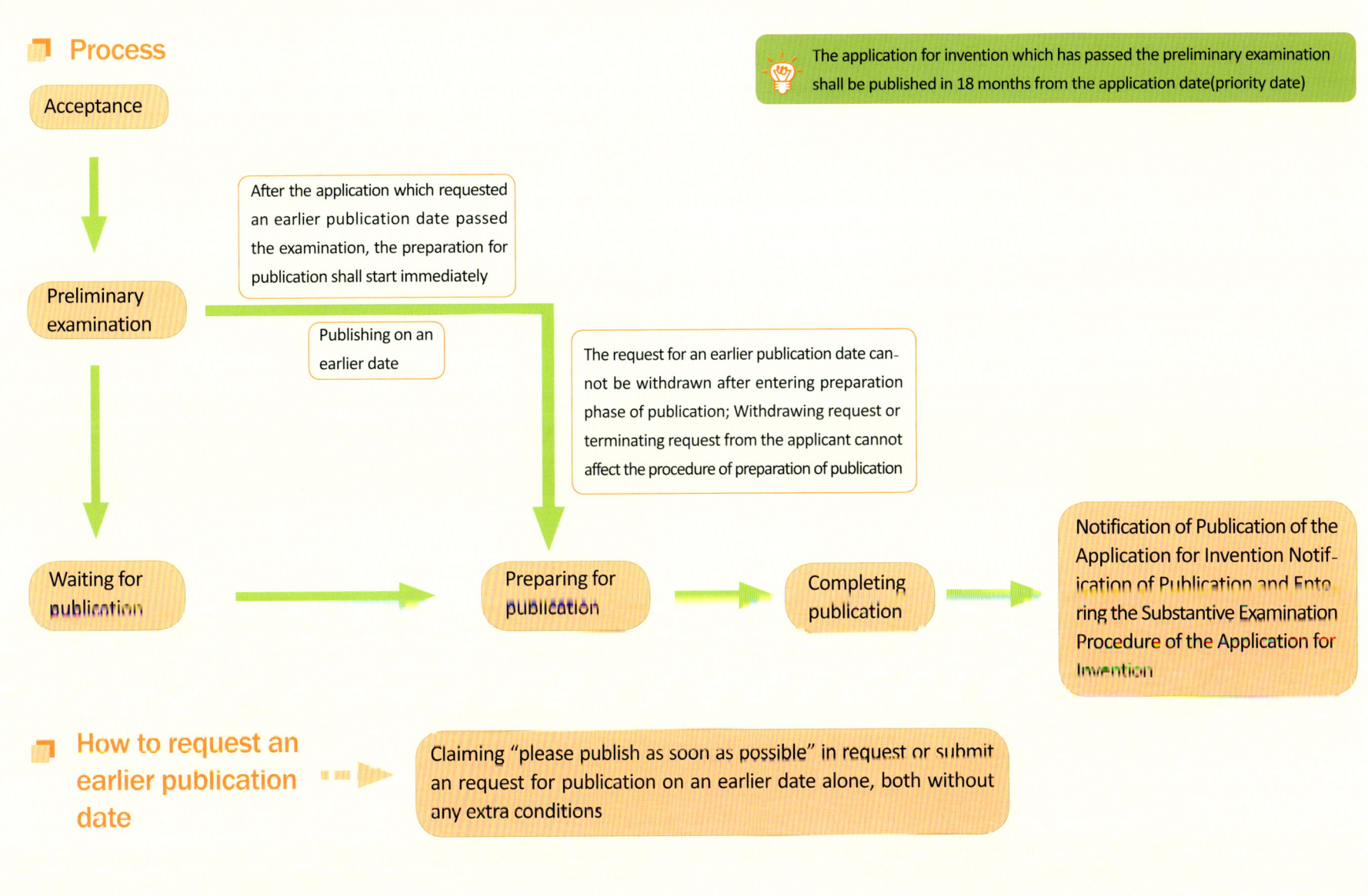

Process

Acceptance

The application for invention which has passed the preliminary examination shall be published in 18 months from the application date(priority date)

After the application which requested an earlier publication date passed the examination, the preparation for publication shall start immediately

Preliminary examination

Publishing on an earlier date

The request for an earlier publication date cannot be withdrawn after entering preparation phase of publication; Withdrawing request or terminating request from the applicant cannot affect the procedure of preparation of publication

Waiting for publication

Preparing for publication

Completing publication

Notification of Publication of the Application for Invention Notification of Publication and Entering the Substantive Examination Procedure of the Application for Invention

How to request an earlier publication date

Claiming "please publish as soon as possible" in request or submit an request for publication on an earlier date alone, both without any extra conditions

办理登记手续

■ **办理登记手续时应缴纳哪些费用**

专利登记费、公告印刷费、授权当年的年费、印花税（按办理登记手续通知书中写明的费用金额缴纳）

4 登记、公告、颁发专利证书

3 进入授权公告准备

2 申请人在期限内办理登记手续，缴纳费用

逾期未办理登记手续

视为放弃取得的专利权

1 专利局发出：办理登记手续通知书、授予专利权通知书

视为放弃取得专利权公告

可在期限内办理恢复手续

逾期未办理恢复手续

■ **办理登记手续的期限** ▶ 自收到办理登记手续通知书之日起2个月内，不可请求延长

Going through Formalities of Registration

🔲 **Fees to pay when going through formalities of registration**

Registration fee, the printing fee for the announcement of grant of patent right, the annual fee of the year in which the patent right is granted, the stamp tax (According to the fee indicated in the Notification of Go through Formalities of Registration)

④ Registration, Announcement, Issuance of Patent Certificate

③ Make preparations for the registration of the grant of the patent right

Entitlement to Patent Deemed Abandoned

② The applicant goes through formalities of registration and pays related fees in the time limit

Fails to go through formalities of registration in time limit

① The Patent Office issues the Notification of Grant Patent Right and the Notification to Go through Formalities of Registration

Announcement of Entitlement to Patent Deemed Abandoned

Going through the formalities of restoration in time limit

Fails to go through the formalities of restoration

🔲 **Time limit for going through formalities of registration**

In 2 months from the date of receiving the Notification of Go through Formalities of Registration, the applicant cannot request to extend the time limit

恢复权利手续

哪些权利丧失可以请求恢复

专利申请权或专利权：
- ◎ 视为撤回
- ◎ 视为放弃取得的专利权
- ◎ 未缴年费专利权终止

其他权利：
- ◎ 视为未要求优先权
- ◎ 生物材料样品视为未保藏
- ◎ 视为未要求不丧失新颖性宽限期

如何办理恢复权利手续

请求恢复的同时，办理权利丧失前应当办理的手续

（1）我有正当理由

（2）收到处分决定起2个月内提出

（3）提交恢复权利请求书

（4）缴纳恢复费1000元

（1）我遇到了不可抗拒的事由

（2）障碍消除后2个月内提出，最迟不超过期限届满起2年

（3）提交恢复权利请求书

（4）提交证明文件

恢复权利请求的审批结果如何

审批通知书

手续不合格：不予恢复

手续合格：准予恢复，继续原程序

Restoration of Right

What can be restored

The right to apply for a patent or patent right:
- Deemed to have been withdrawn
- Deemed to have given up being granted the patent right
- Patent right terminated due to not paying the annual fee

Other rights:
- Deemed not to request for priority
- Biological material sample deemed not to have been deposited
- Grace period concerning novelty deemed not to have been claimed

How to apply

Compensate Processes while applying for reslonation

 (1) I have a warrant

 (2) Submit the warrant within 2 months after receiving the decision

 (3) File the request for resto-ration

 (4) Pay the restoration fee 1000 CNY

 (1) I have irresistible reasons

 (2) Submit the request within 2 months or within 2 years after the expiration date

 (3) File the request for restoration

 (4) Submit the certification documents

Result of restoration

 Notification of examination

 the formality is not qualified : restoration failed

the formality is qualified: restoration approved and continue

57

请求中止程序的手续

何时可以请求中止 ⇒ 因专利申请权或者专利权的归属发生纠纷，已请求管理专利工作的部门调解或者向人民法院起诉的，当事人可以向专利局请求中止有关程序

需要提交哪些材料 ⇒
◎ 中止程序请求书
◎ 管理专利工作的部门或人民法院的写明申请号或专利号的有关受理文件（原件）

审批结果 ⇒
（1）手续合格：中止
（2）手续不合格：视为未提出中止请求

中止何时解除 ⇒ 管理专利工作的部门作出的调解书或者人民法院作出的判决生效后，当事人应当办理恢复有关程序的手续。自请求之日起1年内，相关权属纠纷未能结案的，可以请求延长一次，延长中止期限不超过6个月

中止有关程序的含义是什么 ⇒ 暂停专利申请的审查程序 暂停授予专利权和专利权无效宣告程序 暂停办理权利放弃、权利人变更、专利权质押登记等

Formalities of Request for Suspension

Time to request for suspension

The parties who have a dispute over the right to apply for a patent or ownership of patent right, have requested intellectual property administrative authority for mediation or sued to the People's Court, can request the Patent Office for suspension of relevant procedures

Documents need to submit

- Request for suspension
- The original document for acceptance with patent application number (or patent number)issued by the intellectual property administrative authority or the People's Court

Results after examination and approval

(1)Procedures qualified:suspension

(2)Procedures not qualified:suspension request deemed not to have been made

Cessation of suspension procedure

When the intellectual property administrative authority issues the document of mediation or the judgment made by the People's Court takes effect, the party should go through the formalities for the resumption of the relevant procedures. If no decision is made on the dispute over the ownership right within 1 year of the suspension, the suspension may be extended once, and the extension period shall not exceed 6 months

The significance of suspension of relevant procedures

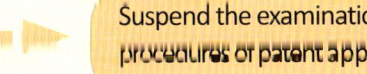 Suspend the examination procedures of patent apply Suspend the procedures of the grant of patent right and invalidation Suspend the formalities to abandon patent right, to make a change of patentee, or to register the pledge of patent right, etc

专利审批流程中请求退款手续

哪些情况可以请求退款 ⇒ 多缴费用：例如应缴年费 600 元，实际缴纳 650 元
重缴费用：例如缴纳 2500 元实审费后再次缴纳
错缴费用：例如缴费时错写专利申请号／专利号

请求退款的时限
自缴费日起 3 年内提交退款请求

谁可提出退款请求 ⇒ 相应费用的缴款人 申请人（非缴款人）：应声明受缴款人委托 ✚ 代理机构（非缴款人）：应声明受缴款人委托

需要提交哪些文件 ⇒ 意见陈述书（关于费用）✚ 缴费收据复印件或汇款证明原件

如何指定退款方式
- ◎ 电子申请的权限人请求退款：电子提交
- ◎ 其他人：可提交纸件

- ◎ 指定通过邮局退款：写明收款人姓名或名称、详细地址及邮编
- ◎ 指定通过银行退款：写明收款人账户名称、账号、开户行(××银行××市××支行)
- ◎ 申请人或代理机构请求退款，未指定退款途径，专利局默认通过邮局退款

The Procedure of Request for Refund in Examination and Approval Flow of Patent Application

In which case can you request a refund

- Overpayments: If the annual fee you shall pay is 600 CNY, the actual fee you paid is 650 CNY
- Duplicate payments: Pay 2500 CNY of substantive examination fee twice
- Wrong payments: Wrongly specify the application number/patent number during payment

Time limit

Within 3 years from the date of payment

Who can request a refund

Remitter of the payment

 Applicant(non-paying party): shall state that he is commissioned by the remitter

 Patent agency(non-paying party): shall state that he is commissioned by the remitter

Materials required

Statement of opinion (about fees)

 The copy of the payment receipt or the original remittance voucher

How to specify the way of refund

- Electronic application user to request: electronic submission
- Others: documents can be submitted in paper form

- Through the post office :indicate the name, detailed address and postal code of the payee
- Through bank : the bank's name(× branch of × bank in × city), the account name and the account number
- Applicant or patent agency request a refund and doesn't specify a refund way: refund through the post office by default

61

专利权评价报告的办理

什么时候可以办理 ➡️ 申请日（优先权日）在 2009 年 10 月 1 日之后的实用新型或外观设计专利，授权公告后

💡 专利权有效或已终止：**可以办理**

💡 专利权已宣告全部无效：**不能办理**

谁可以来办理 ➡️
◎ 全部或部分专利权人，或其代理机构
◎ 利害关系人，或其代理机构

需要提交哪些文件 ➡️

（1）专利权评价报告请求书

（2）专利权人委托新的代理机构或者利害关系人委托代理机构提出请求的，提交委托书并注明委托权限

（3）**独占许可**：被许可人提交独占许可合同或其复印件
普通许可：被许可人提交普通许可合同或其复印件和由专利权人签章的授予起诉权的声明

如何缴纳费用 ➡️ 专利权评价报告请求费 2400 元，自请求之日起 1 个月内缴纳

何时作出评价报告 ➡️ 专利局自收到合格的专利权评价报告请求书和请求费后 2 个月内作出专利权评价报告

如何提出异议 ➡️ 请求人可以在收到评价报告之日起 2 个月内提交意见陈述书提出一次更正请求

How to Process an Evaluation Report of Patent

When ➤
- Patent for utility model or design
- Date of application or priority after October 1,2009
- After the authorization announcement

 Valid or terminated patent：available

 Patent declared invalid in whole：unavailable

Who ➤
- All or part of patentee, or agency
- Interested party or agency

What to submit ➤

(1) Request for Evaluation Report of Patent

(2) Patentee who appointed a new agency or interested party who appointed an agency ,should submit the letter of at‑torney and specify delegation authority

(3) The licensee of exclusive patent license: should submit the exclusive patent license contract or its copy
the licensee of common patent license: should submit the common patent license contract or its copy and the documentation of the authorization of right of action by the patentee

How to pay the fee ➤

Request fee for evaluation report of patent is 2400 CNY, should be paid in one month after submitting the request

When the report can be made ➤

The Patent Office shall make the evaluation report of patent within two months from receiving the eligible request for an Evaluation Report of Patent and the fee for request

How to raise objections ➤

The petitioner may submit a request for reexamination within two months from the date of receipt of the Evaluation Report of Patent for once

邮路查询手续

未收到通知书或专利证书，
且未查询到退信，怎么办

责任在专利局或邮局的，重新发出有关通知和决定或证书

责任在收件人单位收发部门或收件人本人及其有关人员的，
专利局可依请求发送通知书副本或复印件（不变更发文日），
不重新发证书

自发文日起 10 个月内提交
意见陈述书，请求邮路查询

国家知识产权局

邮局向专利局反馈邮路查询结果

邮局

符合邮路查询条件的，专利局向邮局查询

The Formalities of Post Status Inquiry

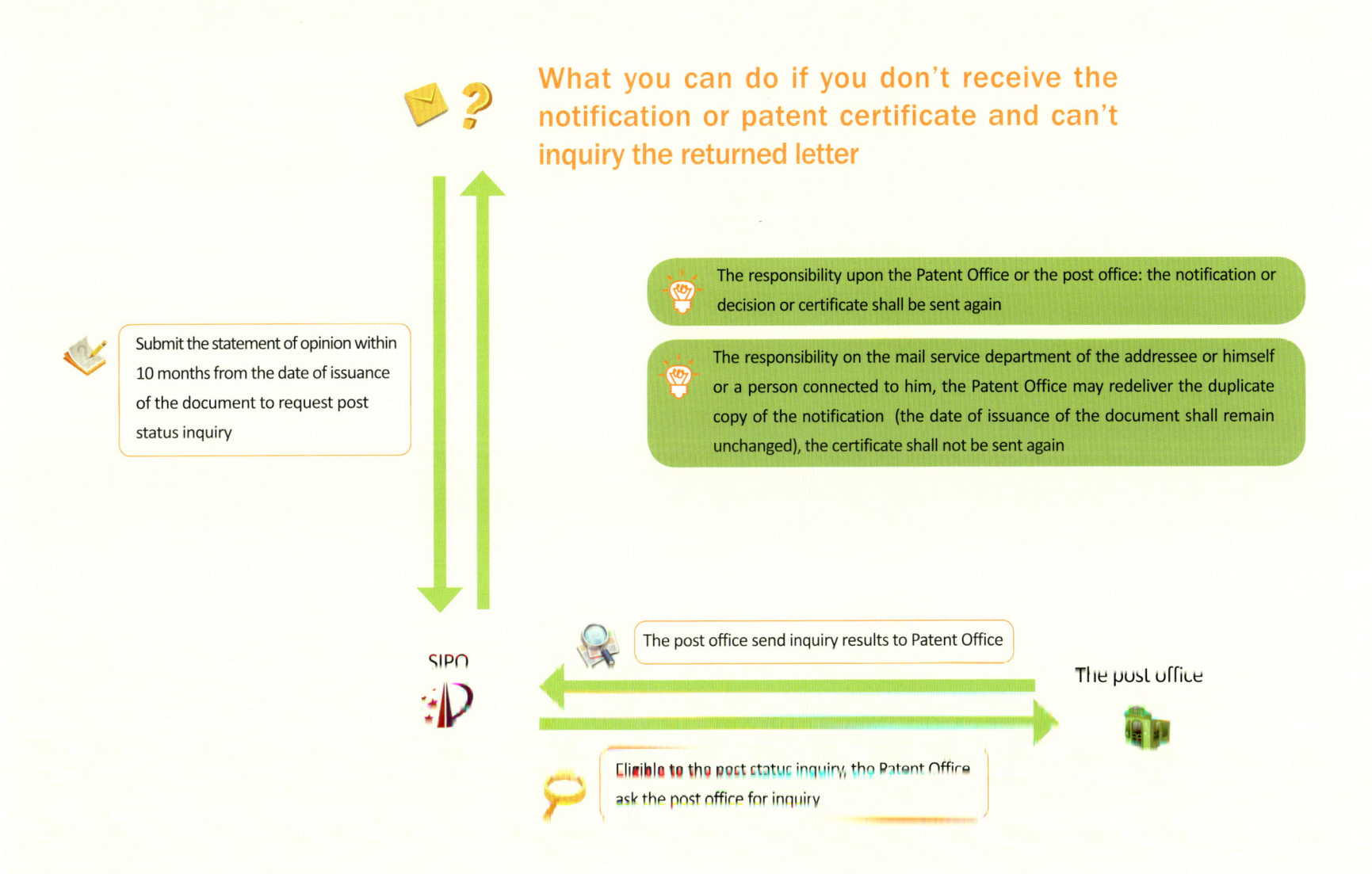

What you can do if you don't receive the notification or patent certificate and can't inquiry the returned letter

The responsibility upon the Patent Office or the post office: the notification or decision or certificate shall be sent again

The responsibility on the mail service department of the addressee or himself or a person connected to him, the Patent Office may redeliver the duplicate copy of the notification (the date of issuance of the document shall remain unchanged), the certificate shall not be sent again

Submit the statement of opinion within 10 months from the date of issuance of the document to request post status inquiry

SIPO

The post office send inquiry results to Patent Office

The post office

Eligible to the post status inquiry, the Patent Office ask the post office for inquiry

副本办理

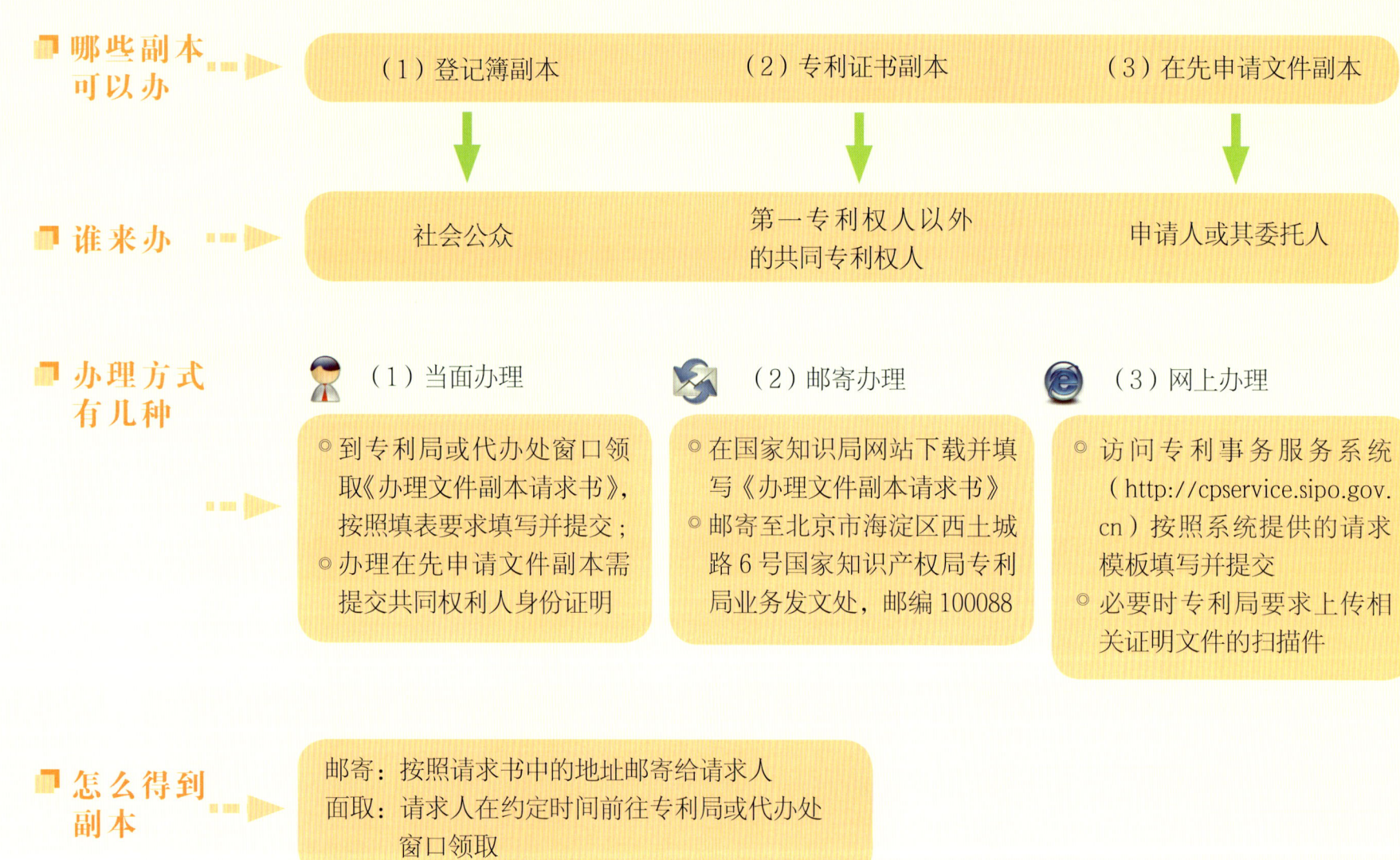

哪些副本可以办

（1）登记簿副本　　　（2）专利证书副本　　　（3）在先申请文件副本

谁来办

社会公众　　　　第一专利权人以外的共同专利权人　　　　申请人或其委托人

办理方式有几种

（1）当面办理
- 到专利局或代办处窗口领取《办理文件副本请求书》，按照填表要求填写并提交；
- 办理在先申请文件副本需提交共同权利人身份证明

（2）邮寄办理
- 在国家知识局网站下载并填写《办理文件副本请求书》
- 邮寄至北京市海淀区西土城路6号国家知识产权局专利局业务发文处，邮编100088

（3）网上办理
- 访问专利事务服务系统（http://cpservice.sipo.gov.cn）按照系统提供的请求模板填写并提交
- 必要时专利局要求上传相关证明文件的扫描件

怎么得到副本

邮寄：按照请求书中的地址邮寄给请求人
面取：请求人在约定时间前往专利局或代办处窗口领取

Requesting Copies

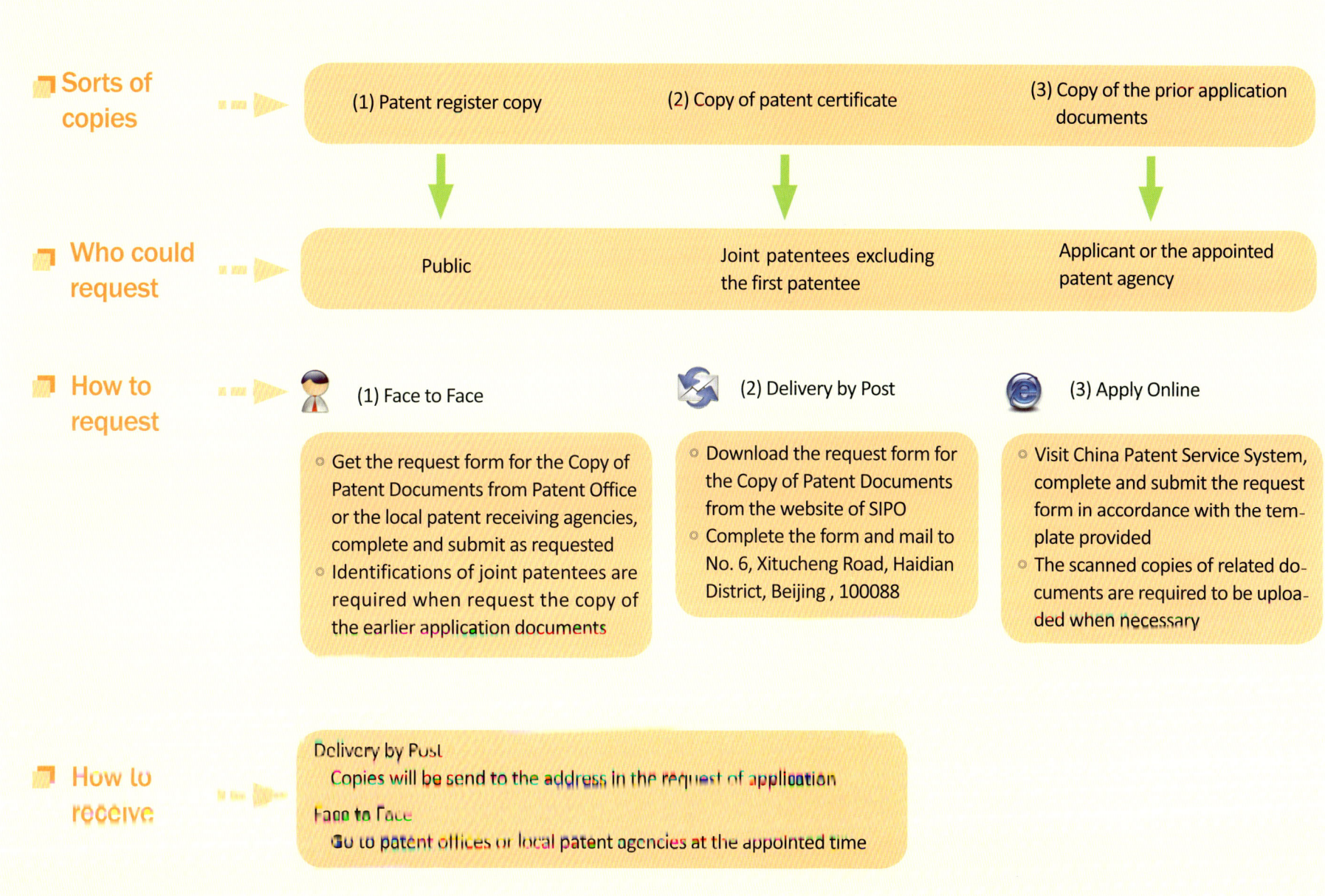

Sorts of copies

(1) Patent register copy

(2) Copy of patent certificate

(3) Copy of the prior application documents

Who could request

Public

Joint patentees excluding the first patentee

Applicant or the appointed patent agency

How to request

(1) Face to Face

- Get the request form for the Copy of Patent Documents from Patent Office or the local patent receiving agencies, complete and submit as requested
- Identifications of joint patentees are required when request the copy of the earlier application documents

(2) Delivery by Post

- Download the request form for the Copy of Patent Documents from the website of SIPO
- Complete the form and mail to No. 6, Xitucheng Road, Haidian District, Beijing , 100088

(3) Apply Online

- Visit China Patent Service System, complete and submit the request form in accordance with the template provided
- The scanned copies of related documents are required to be uploaded when necessary

How to receive

Delivery by Post
Copies will be send to the address in the request of application

Face to Face
Go to patent offices or local patent agencies at the appointed time

专利权质押登记

何时办 ┈▶ 质押合同订立之后　　**谁去办** ┈▶ 出质人和质权人共同办理

需要准备哪些材料 ┈▶
（1）专利权质押登记申请表（当事人或其委托的代理机构签章）
（2）专利权质押合同原件或者经公证机构公证的复印件
（3）出质人和质权人身份证明
（4）出质人、质权人、被委托人共同签章的委托书原件和被委托人身份证复印件

办理方式 ┈▶
（1）当面办理
　　专利局受理大厅 107 房间或地方代办处
　　涉外质押合同只能到专利局登记

（2）邮寄办理
　　邮寄至北京市海淀区西土城路 6 号
　　国家知识产权局专利局业务发文处，
　　邮编 100088，信封注明"质押登记"

不予登记的情形举例 ┈▶
（1）出质人不是合法专利权人或不是全体专利权人　❌
（2）专利权处于年费缴纳滞纳期，或已终止，或被宣告无效　❌
（3）专利权处于中止状态　❌
（4）合同约定债务履行期届满质权人未受清偿时，专利权归质权人所有　❌

Patent Pledge Registration

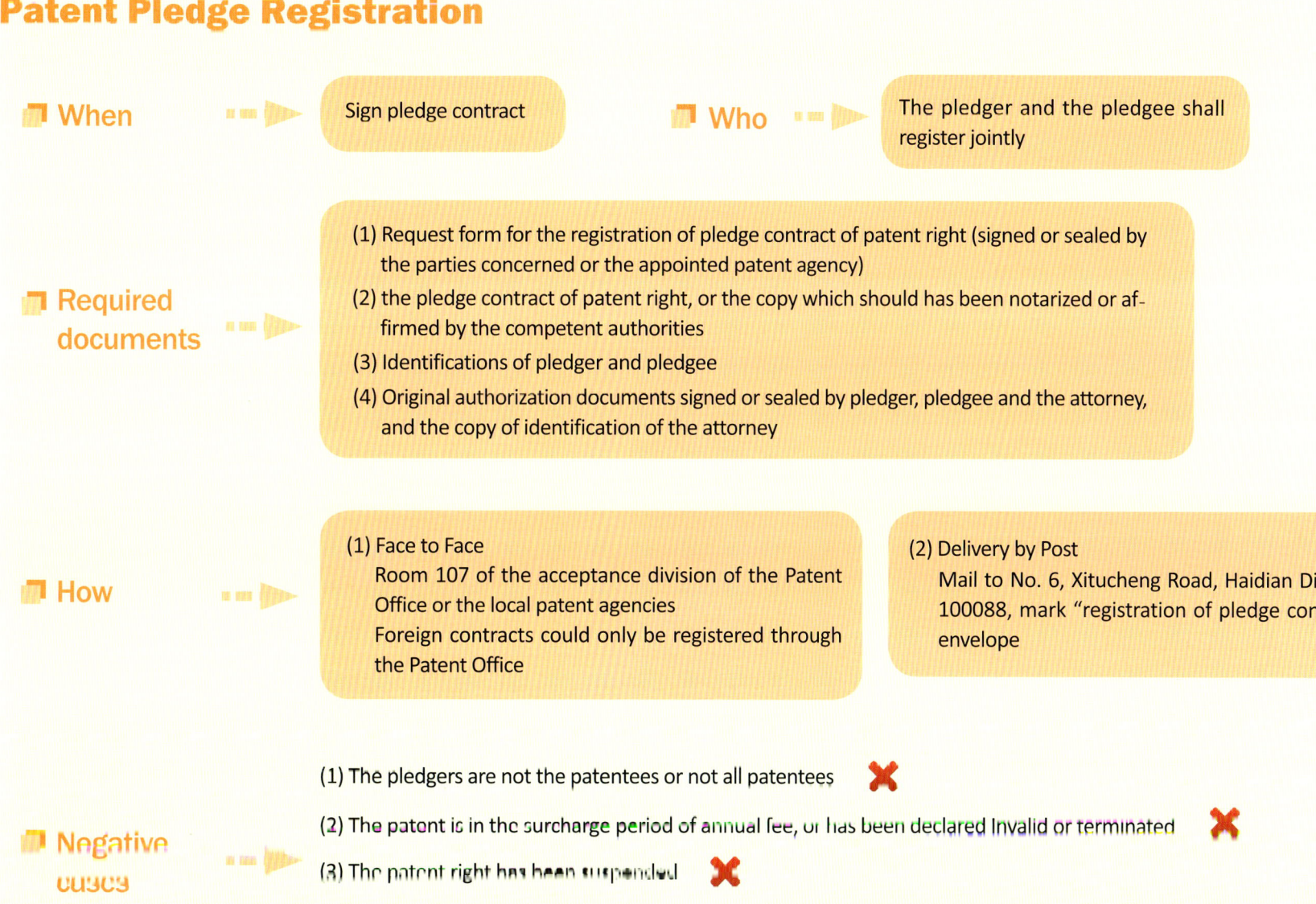

When ➤ Sign pledge contract

Who ➤ The pledger and the pledgee shall register jointly

Required documents ➤
(1) Request form for the registration of pledge contract of patent right (signed or sealed by the parties concerned or the appointed patent agency)

(2) the pledge contract of patent right, or the copy which should has been notarized or affirmed by the competent authorities

(3) Identifications of pledger and pledgee

(4) Original authorization documents signed or sealed by pledger, pledgee and the attorney, and the copy of identification of the attorney

How ➤
(1) Face to Face
Room 107 of the acceptance division of the Patent Office or the local patent agencies
Foreign contracts could only be registered through the Patent Office

(2) Delivery by Post
Mail to No. 6, Xitucheng Road, Haidian District, Beijing, 100088, mark "registration of pledge contract" on the envelope

Negative cases ➤
(1) The pledgers are not the patentees or not all patentees ✖

(2) The patent is in the surcharge period of annual fee, or has been declared Invalid or terminated ✖

(3) The patent right has been suspended ✖

(4) The time limit for debts expires and the pledgee has not received compensation, the ownership of patent shall be attributed to the pledgee ✖

专利实施许可合同备案

🔲 **何时办** ┅▶ 许可合同生效之日起 3 个月内　　🔲 **谁去办** ┅▶ 许可方或受其委托的人

🔲 **需要准备哪些材料** ┅▶
（1）专利实施许可合同备案申请表（许可人或其委托的代理机构签章）
（2）专利实施许可合同原件或者经公证机构公证的复印件
（3）许可人与被许可人身份证明
（4）许可方、被许可方、被委托人共同签章的委托书原件和委托人身份证复印件

🔲 **办理方式** ┅▶
（1）当面办理
　　专利局受理大厅 107 房间或地方代办处涉外许可合同只能到专利局备案

（2）邮寄办理
　　邮寄至北京市海淀区西土城路 6 号国家知识产权局专利局业务发文处，邮编 100088，信封注明"合同备案"

🔲 **不予备案的情形举例** ┅▶

（1）许可人不是合法专利权人或专利申请人或者其他权利人 ❌

（2）专利权被质押且未经质权人同意 ❌

（3）专利权处于年费缴纳滞纳期，或已终止，或被宣告无效 ❌

（4）专利权处于中止状态 ❌

Patent License Registration

 **When** ---▶ In 3 months after the effective date of the license contract

 **Who** ▪▪▶ The licensor or its attorney

 **Required documents** ▪▪▶
(1) Request form for the submission for record of license contract for patent right (signed or sealed by the parties concerned or the appointed patent agency)
(2) Original license contract , or the copy which has been notarized by one notory organization.
(3) Identifications of licensor and licenee
(4) Original documents of authorization signed or sealed by licensor, licenee and the attorney, and the copy of identification of the attorney

 **How to Submit** ▪▪▶

(1) Face to Face
Room 107 of the Acceptance Division of the Patent Office or the local patent agencies
Foreign contracts could only be recorded through the Patent Office

(2) Delivery by Post
Mail to No. 6, Xitucheng Road, Haidian District, Beijing , 100088, mark "submission for record of license contract"on the envelope

Negative Cases ▪▪▶

(1) The licensor is not patentee, patent applicant or any other right holder

(2) The patent has been pledged and the pledgee grant no approval on the license

(3) The patent is in the surcharge period of annual fee, or has been declared invalid or terminated

(4) The patent has been suspended

专利审批流程中的期限

法定期限： 《专利法》及其实施细则规定

期限起点： （1）固定日期（例如申请日）
（2）推定收到通知书之日（发文日加15天）

指定期限： 专利局依部门规章在通知书中指定

期限终点： （1）期限起点＋期限长度
（2）遇休假日顺延

2015 年 2 月						
星期日	星期一	星期二	星期三	星期四	星期五	星期六
1	2	3	4	5	6	7
	假设补正通知书 2014 年 12 月 3 日发文			加 15 天 推定 2014 年 12 月 18 日收到		
						14
15	16	17	18	19	20	21
	加 2 个月长度		除夕 →	春节 →	初二 →	初三 →
22	23	24	25	26	27	28
初四 →	初五 →	初六 →	★	顺延至工作日		

◼ 指定期限的延长

（1）期限届满前提交请求
（2）缴纳延长请求费
（3）可延长 1 个或 2 个月

◼ 常见法定期限

以发明专利为例

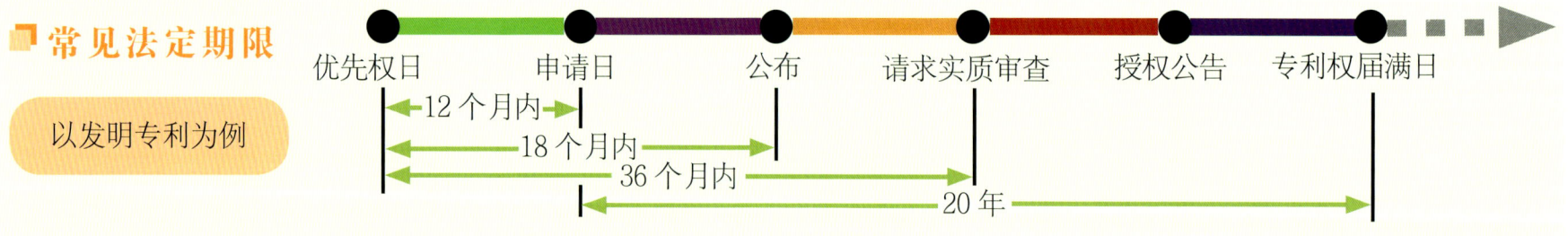

优先权日　申请日　公布　请求实质审查　授权公告　专利权届满日

← 12 个月内 →

← 18 个月内 →

← 36 个月内 →

← 20 年 →

Time Limit

Prescribed Time Limit: Prescribed in the Patent Law and its Implementing Regulations

Specified Time Limit: Specified in notifications issued by patent office

Dies a Quo
(1)Fixed Date(e.g. filing date)
(2)The date on which a notification is presumably received (15 days after the date of the issuance)

Expired Date:
(1)Effective date duration
(2)Postpone when expired on Holiday

Feb, 2015						
SUN	MON	TUE	WED	THU	FRI	SAT
1	2	3	4	5	6	7
Mailing day of rectified notification is Dec.3rd 2014			After 15 days, the presumably received day is Dec.18th 2014			14
15	16	17	18	19	20	21
2 months duration			Holiday	Holiday	Holiday	Holiday
22	23	24	25	26	27	28
Holiday	Holiday	Holiday	Postpone to Working Day			

Extension of Specific Time Limit

(1) Request before the expiration of the time limit
(2) Fee for requesting the extension
(3) Extend 1 or 2months upon request

Prescribed Time Limit

Take Invention as an example

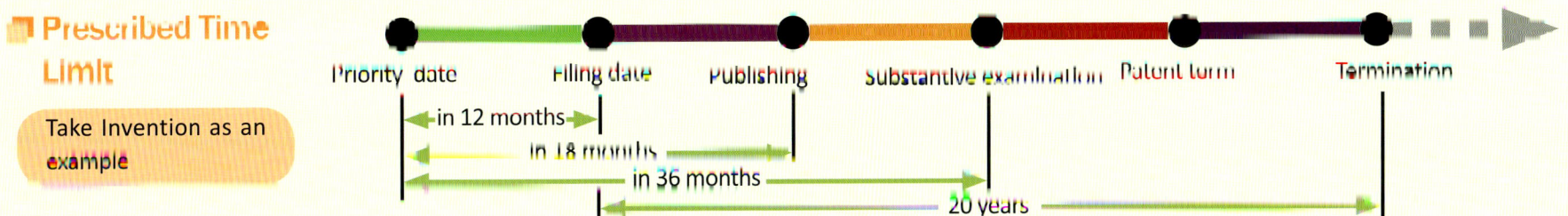

Priority date — Filing date — Publishing — Substantive examination — Patent term — Termination

in 12 months
in 18 months
in 36 months
20 years

集成电路布图设计登记

什么设计可以登记 ┈▶ 芯片 (IC)　　　　电路板（PCB）设计**不可以登记**

申请文件提交方式 ┈▶ 当面提交至专利局 ✚ 邮寄提交至专利局 ✚ 电子提交 电子申请用户登录 http://vlsi.sipo.gov.cn 办理

需要提交哪些文件 ┈▶ 登记申请表 ✚ 图样 ✚ 图样目录 ✚ 委托书（委托代理机构的申请） ✚ 已投入商业利用的设计需提交样品4件

需要缴纳哪些费用 ┈▶ 收到缴费通知书后缴纳：
登记费 2000 元、印花税 5 元，无减缓，无年费

保护期限怎么算 ┈▶ 保护期 10 年，起算日是登记申请日、首次商业利用日，两者较前日期；
无论是否登记或投入商业利用，自创作完成日起 15 年后，不再受保护

Registration of Layout-designs of Integrated Circuits

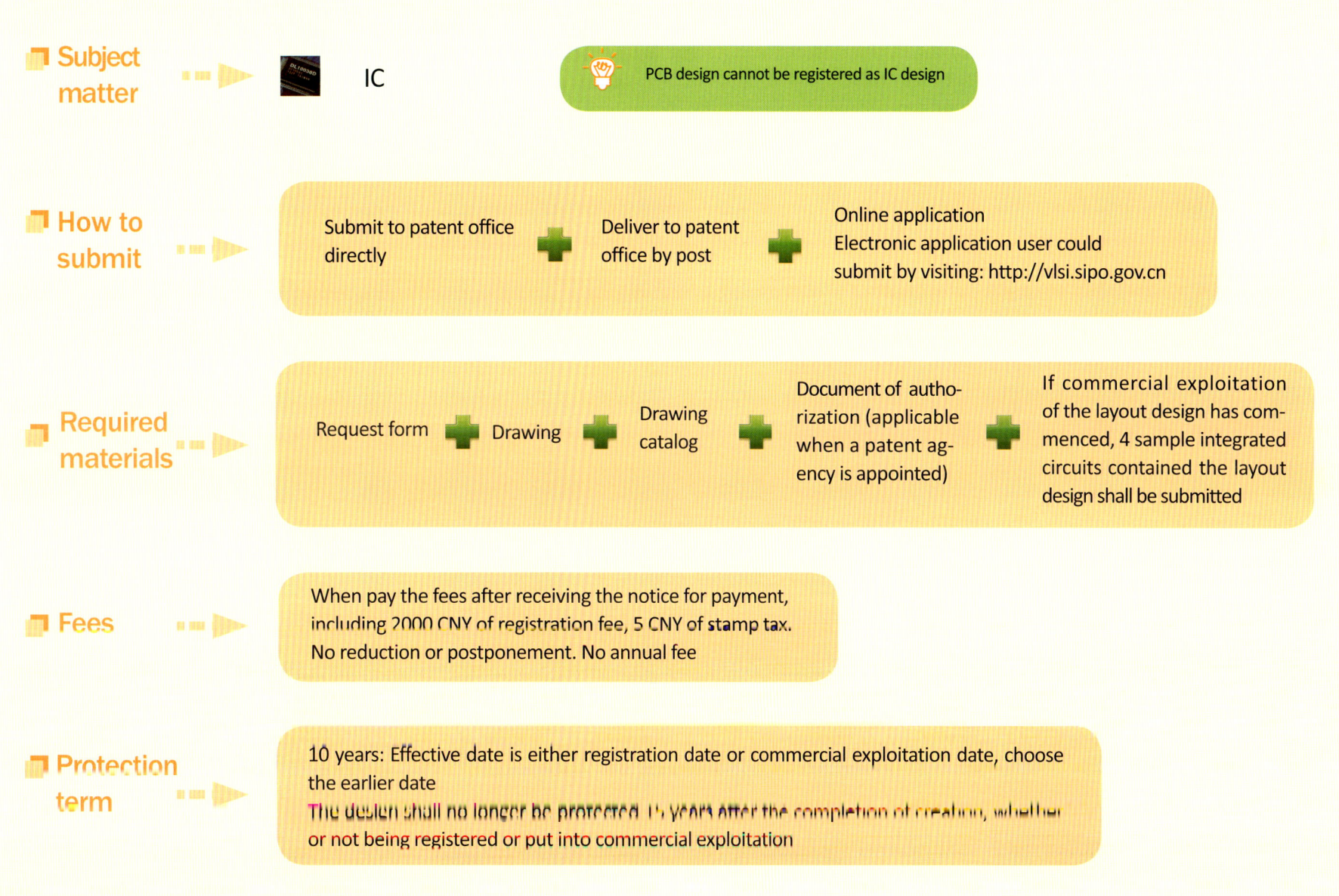

Subject matter ⟶ IC

PCB design cannot be registered as IC design

How to submit ⟶

Submit to patent office directly ➕ Deliver to patent office by post ➕ Online application Electronic application user could submit by visiting: http://vlsi.sipo.gov.cn

Required materials ⟶

Request form ➕ Drawing ➕ Drawing catalog ➕ Document of authorization (applicable when a patent agency is appointed) ➕ If commercial exploitation of the layout design has commenced, 4 sample integrated circuits contained the layout design shall be submitted

Fees ⟶

When pay the fees after receiving the notice for payment, including 2000 CNY of registration fee, 5 CNY of stamp tax. No reduction or postponement. No annual fee

Protection term ⟶

10 years: Effective date is either registration date or commercial exploitation date, choose the earlier date
The design shall no longer be protected 15 years after the completion of creation, whether or not being registered or put into commercial exploitation

与专利审批相关的行政复议程序

对哪些决定不服 可提出行政复议

专利申请不予受理、申请日的确定、保密或不予保密、视为放弃取得专利权、不予恢复专利权、视为撤回、专利权终止……

对驳回决定，复审决定，无效宣告决定，专利局作为国际申请的受理单位、国际检索单位和国际初步审查单位所作决定不服的，不属于行政复议范围

何时可以提出行政复议

自知道具体行政行为之日起60日内可以提出行政复议申请，因不可抗力或者其他正当理由耽误该期限的，该期限自障碍消除之日起继续计算

请求行政复议 提交哪些材料

（1）填写复议申请书，向专利局法律事务处邮寄或递交
（2）附具记载行政决定的通知书复印件
（3）委托专利代理机构或他人的，提交授权委托书
（4）其他证据材料

请求行政复议无须缴费

复议决定

复议决定可能维持、撤销或者变更原行政行为

对复议决定 不服怎么办

可以自收到复议决定之日起15日内向北京知识产权法院起诉

Administrative Reconsideration About Patent Examination and Approval

 Administrative reconsideration shall be implemented on such decisions

Patent application unaccepted/Determination of the date of application/Secrecy or not/Deemed to have abandoned right to obtain the patent right/ Patent right unrecoverable/Deemed withdrawal/ Cessation of patent right

When to apply for administrative reconsideration

The administrative reconsideration application can be put forward since 60 days from the date of the specific administrative act, if the time limit is to be delayed due to force majeure or any other justifiable reason, it will continue to be calculated from the date of the elimination of the obstacle

Dissatisfied to the decision of rejection/reexamination/invalidation or the decision made by the Patent Office as agency for international applications, International research and international preliminary examination, do not belong to the scope of administrative reconsideration

Materials to apply for administrative reconsideration

(1) Fill in the application for reconsideration, mail or submit to office of legal affairs of the Patent Office
(2) Attach the copy of notice of administrative decision
(3) Submit the authorization documents if you have appointed patent agency or other people
(4) Other evidence materials

Free of charge

Decision of Reconsideration

The decision of reconsideration may stand still, withdraw or change original administrative act

Dissatisfied to the decision of administrative reconsideration

You can begin a suit in Beijing intellectual property court within 15 days from the date of receipt of the reconsideration decision